CONSOLIDATED LIST
OF SCIENTIFIC AND TECHNICAL SERIALS

CONSOLIDATED LIST
OF SCIENTIFIC AND TECHNICAL SERIALS
REGULARLY SEEN BY THE INSTITUTES AND BUREAUX
OF THE
COMMONWEALTH AGRICULTURAL BUREAUX

COMMONWEALTH AGRICULTURAL BUREAUX

First published in 1971
by the
Commonwealth Agricultural Bureaux.
Farnham Royal, Slough SL2 3BN, England.
Price £6 - 00

This and other publications of the Commonwealth
Agricultural Bureaux. can be obtained through any
major bookseller or direct from:

COMMONWEALTH AGRICULTURAL BUREAUX
Central Sales, Farnham Royal,
Slough SL2 3BN, England.

PRINTED IN GREAT BRITAIN

PREFACE

Lists of serials regularly seen by each CAB Institute
and Bureau are published separately, approximately
every two years.   The lists have been consolidated
and the following pages contain, in alphabetical order,
the titles of all serials regularly seen.

1970

A

A.E.A. Information Series.
Louisiana Agricultural Experiment
Station, Baton Rouge. LA.

A.E.C. Atomic Energy Commission.
Washington

A.I. Digest. Columbia

A.I.B.S. Bulletin. Washington

A.I.C.C. Economic Review, New Delhi

A.P.A. Quarterly: American Society
for Testing Materials.
Philadelphia

A.P.P.I.T.A. Melbourne

A.S.L.I.B. Book List. London

A.S.L.I.B. Proceedings. London

Abhandlungen und Berichte des
Naturkundesmuseums-Forschungs-
stelle. Gorlitz.

Abhandlungen und Berichte aus dem
Staatlichen Museum fur Mineralogie und Geologie. Dresden

Abhandlungen der Braunschweigi-
schen wissenschaftlichen
Gesellschaft.

Abhandlungen der Deutschen
Akademie der Wissenschaften zu
Berlin

  Klasse fur Chemie, Geologie,
  Biologie.

Abhandlungen hrsg. von der Sencken-
bergischen naturforschenden
Gesellschaft. Frankfurt a.M.

Abhandlungen und Verhandlungen des
Naturwissenschaftlichen Vereins
in Hamburg

Abstracts of Bulgarian Scientific
Literature. Sofia.

  Agriculture and Forestry
  Veterinary Medicine

Abstracts of Dissertations.
University of Cambridge

Abstracts of Dissertations.
University of Maryland. USA

Abstracts of Doctoral Dissertations
Ohio State University. Columbus

Abstracts of Doctoral Dissertations
University of Nebraska. Lincoln

Abstracts on Hygiene. UK

Abstracts of the Meeting of the
Japan Gibberellin Research
Association. Tokyo

Abstracts of the Meeting of the
Weed Society of America

Abstracts of Mycology.
Philadelphia

Abstracts of Papers. American
Chemical Society. Washington

Abstracts of Papers. American
Society for Horticultural
Science. Ithaca

Abstracts of Published Papers and
List of Translations. C.S.I.R.O.
Australia. Melbourne

Abstracts of Rumanian Technical
Literature

Academia Nacional de Agronomia y
Veterinaria. Buenos Aires

Academie de la Republique
Populaire Roumaine. Revue de
Physique. Bucharest

Acarologia. Abbeville, Paris

Accademia medica. Genova, Torino

Accounts of Chemical Research.
Washington

Achievements of the Soil Science in
the Ukraine. Reports to the 8th
International Congress of Soil
Science. Kiev.

Acqua nell' agricoltura, nell'
igiene e nell' industria. Roma

Acridological Abstracts. London

Acta agralia fennica. Helsinki

Acta agraria et silvestria. Poland

  Seria Silvestris
  Seria Polnicza
  Seria Zootechniczna

Acta agriculturae scandinavica.
Stockholm

Acta agriculturae sinica. Peking

Acta agrobotanica. Warszawa

Acta agronomica. Palmira

Acta agronomica Academiae Scientiarum Hungaricae. Budapest

Acta allergologica. Kóbenhavn.

Acta anatomica. Basel, New York

Acta Biochemica et Biophysica Academiae Scientiarum Hungarican. Budapest

Acta biochimica polonica. Warszawa

Acta biochimica sinica. Peking.

Acta biologiae experimentalis sinica. Peking

Acta biologica Academiae Scientiarum Hungaricae. Budapest

Acta biologica Academiae Scientiarum Hungaricae (Supplement) Budapest

Acta biologica Cracoviensia Seri. Zoology

Acta biologica Jugoslavica Ser.B.

Acta biologica et medica germanica. Berlin

Acta Biologica venezuelica. Caracas

Acta botanica Academiae Scientiarum Hungaricae. Budapest

Acta botanica croatica. Zagreb

Acta botanica neerlandica. Amsterdam

Acta botanica sinica. Peking

Acta botanica venezuelica

Acta chemica scandinavica Kóbenhavn

Acta chirurgica italica. Padova

Acta chirurgica scandinavica. Stockholm

Acta cientifica potosina. San Luis Potosi

Acta cientifica venezolana. Caracas

Acta dermato-venereologica. Stockholm

Acta embryologiae et morphologiae experimentalis. Palermo

Acta endocrinologica. Copenhagen

Acta entomologica bohemoslovaca. Czech.

Acta entomologica fennica. Helsinki

Acta entomologica Musei nationalis. Czech.

Acta entomologica sinica. Peking

Acta ethnographica Academiae Scientiarum Hungaricae. Budapest

Acta faunistica entomologica Musei nationalis Pragae.

Acta forestalia fennica. Helsingforsiae.

Acta fytotecnica. Nitra

Acta gastro-enterologica belgica. Bruxelles

Acta genetica et statistica medica. Basle

Acta geographica sinica. Peking

Acta geologica Academiae Scientiarum Hungaricae. Budapest

Acta gerontologica. Milano

Acta haematologica. Basel, New York

Acta haematologica japonica. Kyoto

Acta histochemica. Jena

Acta horticulturalia sinica. Peking

Acta hydrobiologica. Kraków

Acta medica. Fukuoka

Acta medica Academiae Scientiarum Hungaricae. Budapest

Acta medica et biologica. Niigata

Acta medica costarricense. San José.

Acta medica italica di malattie infettive e parassitarie. Roma

Acta medica jugoslavica. Beograd

Acta medica orientalia. Jerusalem, Tel Aviv

Acta medica philippina. Manila

Acta medica scandinavica. Stockholm

Acta medica scandinavica (Supplements). Stockholm

2

Acta medica turcica. Ankara

Acta medica veterinaria. Napoli

Acta medicinae Okoyama. Okoyama

Acta microbiologica Academiae
Scientiarum Hungaricae.
Budapest

Acta microbiologica hellenica.

Acta microbiologica polonica.
Warszawa

Acta microbiologica sinica. Peking

Acta morphologica Academiae
Scientiarum Hungaricae. Budapest

Acta morphologica neerlando-
scandinavica. Utrecht.

Acta Musei macadonici scientiarum
naturalium. Skopje.

Acta Musei Moraviae (Scientiae
naturales). Czech.

Acta Mycologica. Polskie
Towarzystwo Botaniczne. Warsaw

Acta neurochirurgica. Wien

Acta neurologica et psychiatrica
belgica. Bruxelles

Acta neuropathologica. Berlin

Acta obstetrica et gynecologica
scandinavica. Helsingfors, etc.

Acta Obstetrica et gynecologica
scandinavica. Supplements.
Lund

Acta Oeconomica. Budapest

Acta operativo-aeconomica
Universitatis Agriculturae.
Nitra

Acta operativo-oeconomica
Universitatis Agriculturae.
Nitra

Acta ophthalmologica. Kjøbenhavn

Acta orthopaedica scandinavica.
Kjøbenhavn

Acta orthopaedica scandinavica.
(Supplement). Kjøbenhavn

Acta Paediatrica Academiae
Scientiarum Hungaricae. Budapest

Acta paediatrica belgica.
Bruxelles

Acta Paediatrica Japonica
(Domestic Edn.) Tokyo

Acta Paediatrica Japonica
(Overseas Edn.) Tokyo

Acta Paediatrica Scandinavica.
Uppsala

Acta Paediatrica Scandinavica.
Supplements. Stockholm

Acta palaeobotanica. Cracovia

Acta parasitologica lithuanica.
Vilnius.

Acta parasitologica polonica.
Warszawa

Acta pathologica japonica. Tokyo

Acta pathologica et microbiologica
scandinavica. Kjøbenhavn

Acta pathologica et microbiologica
scandinavica. (Supplement)
Kjøbenhavn

Acta pediatrica española. Madrid

Acta pedologica sinica. Nanking

Acta pharmaceutica jugoslavica.
Zagreb

Acta pharmaceutica sinica. Peking

Acta pharmacologica et toxicologica
Kjøbenhavn

Acta physiologica Academiae Scientiarum Hungaricae. Budapest

Acta physiologica latinoamericana.
Buenos Aires

Acta physiologica et pharmacologica
néerlandica. Amsterdam, etc.

Acta physiologica polonica.
Warszawa

Acta physiologica scandinavica.
Stockholm

Acta physiologica scandinavica.
Supplements. Stockholm

Acta physiologica sinica. Peking

Acta phytogeographica suecica.
Uppsala

Acta phytopathologica Academiae
Scientiarum Hungaricae. Budapest

Acta phytopathologica sinica.
Peking

Acta phytophylacica sinica.
Peking

Acta phytotaxonomica et geobotanica
Kyoto

Acta phytotaxonomica sinica.
Peking

Acta protozoologica. Poland

Acta pruhoniciana. Pruhonice.

Acta radiologica. Stockholm

Diagnosis
Therapy, Physics, Biology

Acta Scholae medicinalis in Gifu.
Gifu

Acta Scholae medicinalis Universitatis in Kioto. Kioto

Acta Societatis botanicorum
Poloniae. Warszawa.

Acta salmanticensia. Salamanca.

Serie de Ciencias

Acta Societatis medicorum upsaliensis. Uppsala

Acta Societatis zoologicae
bohemoslovenicae. Czech.

Acta Technologica Agriculturae
Universitatis Agriculturae.
Nitra

Acta theriologica. Warszawa

Acta tropica. Basel

Acta Universitatis Agriculturae.
Facultas Agroeconomica. Brno

Acta Universitatis Agriculturae.
Facultas Agronomica. Brno

Acta Universitatis Agriculturae
et Silviculturae. Brno

Acta Universitatis Carolinae.
Prague

Biologica
Medica

Acta Universitatis lundensis.
Lund

Acta Veterinaria. Beograd

Acta Veterinaria. Brno
(Formerly: Acta Univ.Agric.Fac.
Vet.)

Acta veterinaria Academiae Scientiarum Hungaricae. Budapest

Acta veterinaria japonica. Tokyo

Acta veterinaria scandinavica.
Copenhagen

Acta veterinaria et zootechnica
sinica. Peking

Acta virologica. Prague

Acta vitaminologica et
enzymologica

Acta zoologica. Stockholm

Acta zoologica Academiae Scientiarum Hungaricae. Budapest

Acta zoologica cracoviensia.
Kraków.

Acta zoologica fennica.
Helsinforsiae.

Acta zoologica lilloana. Tucumán.

Acta zoologica mexicana. Mexico

Acta zoologica et oecologica
Universitatis lodziensis. Lódź

Acta zoologica et pathologica
antverpiensia. Belgium

Acta zoologica sinica. Peking

Acta zootechnica. Nitra.

Actas dermo-sifilograficas. Madrid

Actas de la Reunion Nacional de la
Sociedad española de Ciencias
Fisiologicas.

Activiteitsverslag. Centrum voor
Landbouwkundig Onderzoek Gent.

Adansonia. Paris

Administration Report of the
Director of Agriculture, Ceylon

Administration Report of the
Director of Agriculture, Trinidad
and Tobago

Administration Report, Forest
Department Andhra Pradesh.
Hyderabad.

Administration Report, Forest
Department Bombay. Poona

Administration Report, Forest
Department Madras.

Advancement of Science. London

Advances in Acarology. Ithaca

Advances in Agronomy. New York

Advances in Applied Microbiology.
New York

Advances in Biological and Medical
Physics. New York

Advances in Botanical Research.
New York

Advances in Carbohydrate Chemistry.
New York

Advances in Chemistry Series.
Washington

Advances in Chromatography.
New York

Advances in Ecological Research

Advances in Enzyme Regulation.
New York

Advances in Enzymology and Related
Subjects of Biochemistry.
New York

Advances in Food Research.
New York

Advances in Genetics. New York

Advances in Immunology. New York

Advances in Insect Physiology.
New York

Advances in Lipid Research.
New York

Advances in Metabolic Disorders.
New York

Advances in Organic Geochemistry.
Oxford

Advances in Parasitology. USA & UK

Advances in Pest Control Research.
New York, London

Advances in Protein Chemistry.
New York

Advances in Reproductive Physiology
UK

Advances in Veterinary Science.
New York

Advances in Virus Research.
New York

Advancing Frontiers of Plant
Sciences. New Delhi.

Advisory Circular. Rubber Research
Institute of Ceylon. Dartonfield

Advisory Leaflet. Department of
Agriculture, Scotland.
Edinburgh

Advisory Leaflet. Division of
Plant Industry, Department of
Agriculture and Stock, Queensland
Brisbane

Advisory Leaflet. Ministry of
Agriculture, Fisheries and
Food. London

Advisory Leaflet. West of
Scotland Agricultural College.
Glasgow.

Advisory Service Leaflet. Timber
Research and Development
Association. (TRADA) High Wycombe

Aerospace Medicine. Washington

Ärztliche Forschung. Bad Wörishafen

Ärztliche Wochenschrift. Berlin

Africa. London

Africa and Irrigation. Proceedings
of an International Symposium.
Salisbury

African Affairs. Oxford

African Report. Washington

African Soils. Paris

African Violet Magazine.
Knoxville

African World. London

Afrika heute. Bonn

Afrika Studien. IFO-Institut für
Wirtschaftsforschung. München

Afrique française chirurgicale.
Alger.

Agra-Europe. Presse- und
Information-dienst. Brussel

Agra University Journal of
Research: Science. Agra.

Agrargazdasági Kutato Intézet
Fütezei. Budapest

Agrarische Rundschau. Wien

Agrarpolitik und Marktwesen.
Hamburg.

Agrarpolitische Revue. Bern

Agrarstatistik. Brussel

Agrártudományi egyetem Központi
könyvtárának kiadványai.
Budapest.

Agrártudományi egyetem Mezőgazdas-
agtudományi karának közlemenyei.
Gödöllő.

Agrártudomány. Budapest

Agrarwirtschaft. Hannover

Agrekon. Pretoria.

Agri Digest. Bruxelles.

Agri Forum. Munchen

Agri hortique genetica. Landskrona.

Agricoltore. Bresciano

Agricoltore d'Italia. Roma

Agricoltura: attualitá italiana e
straniere. Roma.

Agricoltura italiana. Pisa

Agricoltura siciliana.

Agricultura tecnica. Santiago

Agricoltura delle Venezie.
Venezia

Agricultura. Lisboa

Agricultura. Louvain

Agricultura. Madrid

Agricultura em São Paulo

Agricultura y ganaderia. Santiago
de Chile

Agricultura técnica. Santiago de
Chile

Agricultura técnica en México.
Mexico

Agricultura tropical. Bogota

Agricultura Aviation. The Hague

Agricultural and Biological
Chemistry. Tokyo

Agricultural Bulletin. Bermuda
Department of Agriculture.
Hamilton

Agricultural Bulletin. Canterbury
Chamber of Commerce. Christchurch

Agricultural Bulletin of the Saga
University. Saga

Agricultural Chemicals. Baltimore

Agricultural Chemicals Digest. USA

Agricultural Cooperative Monthly
Survey. Seoul

Agricultural Development Paper.
Agricultural Division, FAO.
Washington

Agricultural Development Study
Report. Department of Agricul-
tural Economics. Wye College.
Ashford

Agricultural Economics Bulletin for
Africa. Addis Ababa.

Agricultural Economic Report.
Department of Agricultural
Economics. Michigan State Univ.

Agricultural Economics Report.
Department of Land Economy.
University of Cambridge.

Agricultural Economic Report.
Economic Research Service,
US Department of Agriculture.
Washington

Agricultural Economic Report.
Hawaii Agricultural Experiment
Station. Honolulu.

Agricultural Economic Report.
US Department of Agriculture.
Washington

Agricultural Economics Research.
Washington

Agricultural Economic Research
Bulletin. Department of Extension
University of Alberta. Edmonton

Agricultural Economic Research
Report. University of Western
Australia. Press. Nedlands

Agricultural Economic Review.
Thessaloniki

Agricultural Economic Statistical
Series. North Dakota State Univ.
Agricultural Applied Science.
Fargo

Agricultural Engineering.
St. Joseph, Mich.

Agricultural Finance Review.
Washington

Agricultural Gazette of New South
  Wales. Sydney

Agricultural Handbook. US Department of Agriculture. Washington

Agricultural History. Washington

Agricultural History Review.
  London

Agricultural Information Bulletin.
  Economic Research Service.
  US Department of Agriculture

Agricultural Institute Review.
  Ottawa

Agricultural Ireland. Dublin

Agricultural Journal. Department
  of Agriculture, Fiji. Suva.

Agricultural Literature of
  Czechoslovakia. Prague

Agricultural Magazine of the
  Annamalai University

Agricultural Marketing. USA
Agricultural Merchant. London

Agricultural Meteorology. Amsterdam

Agricultural News Letter. USA

Agricultural Planning Studies.
  FAO Rome

Agricultural Progress. London, etc.

Agricultural Record. Dublin

Agricultural Research. Department
  of Agricultural Technical
  Services, Republic of South
  Africa. Pretoria.

Agricultural Research. Washington

Agricultural Research. Taiwan
  Agricultural Research Institute.
  Taipei

Agricultural Research Journal of
  Kerala. Vellayani

Agricultural Research Report.
  Centre for Agricultural
  Publications and Documentation.
  Wageningen

Agricultural Research Reports
  (Verslagen van Landbouwkundige
  Onderzoekingen). Wageningen

Agricultural Research Review.
  Cairo

Agricultural Science. Hong Kong

Agricultural Science Digest.
  Purdue

Agricultural Science Review.
  Washington

Agricultural Science in South
  Africa. Pretoria. Sect. 3.

Agricultural Series. University of
  West Indies. Barbados

Agricultural Series Leaflet.
  Food and Agriculture Council.
  Pakistan

Agricultural Situation in
  India. Delhi

Agricultural Statistics. Ministry
  of Agriculture, Fisheries and
  Food. London

Agricultural Statistics. Departmen
  of Agriculture for Scotland.
  Edinburgh

Agricultural Statistics.
  US Department of Agriculture.
  Washington

Agricultural Studies. FAO

Agricultural and Veterinary
  Chemicals. London

Agriculture. London

Agriculture. Paris

Agriculture. Quebec.

Agriculture abroad. Ottawa.

Agriculture and Agro-Industries
  Journal. Bombay

Agriculture and Animal Husbandry.
  Uttar Pradesh. Lucknow

Agriculture and Horticulture.
  Tokyo

Agriculture Handbook.
  US Department of Agriculture

Agriculture Information Bulletin.
  Washington

Agriculture Monograph of the
  US Department of Agriculture.
  Washington

Agriculture in Northern Ireland.
  Belfast

Agriculture Pakistan. Karachi

Agriculture Pratique. Paris

Agrikultura. Bratislava

Agrobotanika. Budapest

Agrochimica. Pisa

Agrociencia. Mexico

Agroforum. Berlin

Agrokémia és talajtan. Budapest

Agrokhimiya. Moskva.

Agronomia. Brazil

Agronomia. Caracas

Agronomia. Lima

Agronomia. Manizales

Agronomia. Escola nacional de
agronomia. Rio de Janeiro

Agronomia. Facultad de Agronomia.
Universidad de San Carlos de
Guatemala

Agronomia angolana. Luanda

Agronomia lusitana. Sacavem, etc.

Agronomia Mocambicana. Laurenço
Marques.

Agronomia Monterrey. Mexico

Agronomia tropical. Maracay

Agronomia tropical. Venezuela

Agronomía y veterinaria técnica y
practica rural. Buenos Aires.

Agronômico. Campinas

Agronomie tropicale. Nogent-sur-
Marne, etc.

Agronomski glasnik. Zagreb.

Agronomy. Current Literature
US Department of Agriculture
Washington

Agronomy Journal. Washington

Agros. Lisboa

Agrotecnia de Cuba. Havana

Agrotike Oikonomia. Athenai

A'in Shams Medical Journal. Cairo

Air Pollution Abstracts.
Warren Spring Laboratory

Air and Water Pollution. Oxford

Ajia Keizai. Tokyo

Akola Agricultural College
Magazine.

Akusherstvo i ginekologiya.
Leningrad, Moskva.

Albrecht von Graefes Archiv fur
Ophthalmologie. Leipzig

Albrecht-Thaer-Archiv. Berlin

Alexandria Journal of Agricultural
Research. Alexandria.

Alexandria Medical Journal.
Alexandria.

Algemeen zuivelblad. The Hague

Algérie agricole. Alger

Algerie medicale. Alger

Aligarh Muslim University
Publications. Zoological Series

Aliso. Anaheim, Claremont

Alimenta. Zurich

Alimentation et la Vie. Paris

Alimentazione animale. Roma

Allahabad Farmer. Allahabad

Allam es Igazgatás. Budapest

Allami Gazdaság. Budapest

Allattenyésztés. Budapest

Allgemeine Fischereizeitung.
München

Allgemeine Forstzeitschrift.
München

Allgemeine Forstzeitung. Wien

Allgemeine Forst- und Jagdzeitung.
Frankfurt a.M.

Allionia. Bollettino dell'Istituto
ed orto botanico dell'Universita
di Torino. Torino

Almanaque del Ministerio de
agricultura y ganaderia de la
nacion. Buenos Aires

Alvsborg läns norra hushållnings-
sällskaps tidskrift.

Amaryllis Year Book. La Jolla

Amatus lusitanus. Lisboa

America latina. Rio de Janeiro

American Anthropologist.
Lancaster, etc.

American Camellia Society Yearbook

American Dairy Review. New York

American Economic Review. Menasha

American Fern Journal. Port
Richmond

American Forests. Washington

American Fruit Grower. Willoughby

American Fruit Grower Magazine.
Cleveland

American Fur Breeder. St.Peter.

American Heart Journal. St.Louis

American Horticultural Magazine.
Washington

American Horticultural Society
Gardeners Forum. Washington

American Institute of Crop
Ecology. Washington.

American Journal of Agricultural
Economics. Urbana. Ill.

American Journal of Anatomy.
Baltimore

American Journal of Botany.
Lancaster

American Journal of Cardiology.
New York

American Journal of Clinical
Nutrition. New York. etc.

American Journal of Clinical
Pathology. Baltimore

American Journal of Comparative
Law. Ann Arbor. Mich.

American Journal of Digestive
Diseases. Fort Wayne, New York

American Journal of Diseases of
Children. Chicago

American Journal of Economics and
Sociology. New York

American Journal of Enology and
Viticulture. Delano, Davis.

American Journal of Epidemiology.
USA.

American Journal of Gastroenterology. New York

American Journal of Human Genetics.
Baltimore

American Journal of the Medical
Sciences. Philadelphia

American Journal of Medical
Technology. Detroit. etc.

American Journal of Medicine.
New York

American Journal of Nursing.
Philadelphia

American Journal of Obstetrics and
Gynecology. St. Louis

American Journal of Ophthalmology.
St. Louis

American Journal of Orthopsychiatry
Menasha. etc.

American Journal of Pathology.
Boston.

American Journal of Physical
Anthropology. Washington

American Journal of Physiology.
Boston, etc.

American Journal of Public Health
/and Nation's Health.7 New York

American Journal of Roentgenology,
Radium Therapy and Nuclear
Medicine. New York

American Journal of Science.
New Haven

American Journal of Surgery.
New York

American Journal of Tropical
Medicine and Hygiene. Baltimore

American Journal of Veterinary
Clinical Pathology

American Journal of Veterinary
Research. Chicago

American Midland Naturalist.
Notre Dame

American Naturalist. Lancaster

American Mineralogist. Lancaster

American Nurseryman. Rochester

American Orchid Society Bulletin.

American Orchid Society Yearbook

American Perfumer and Cosmetics.
Pontiac. Ill.

American Potato Journal.
Washington

American Potato Yearbook.
New York

American Poultry Journal. Chicago

American Review of Respiratory
Diseases. Baltimore

American Rose Annual. Harrisburg

American Society of Civil
Engineers. Irrigation and
Drainage Division Journal.
New York

American Statistician. Washington

American Tomato Yearbook.
Westfield

American Vegetable Grower.
Willoughby

American Zoologist.

Amtliche Pflanzenschutzbestim-
mungen. Berlin

Anadolu kliniĝi. Istanbul

Anais da Academia brasileira
de ciencias. Rio de Janeiro

Anais brasileiros de dermatologia
e sifilografia. Rio de Janeiro

Anais da Escola superior de
agricultura "Luiz de Queiroz".
Piracicaba

Anais de Escola superior de
medicina veterinaria. Lisboa

Anais da Faculdade de ciências
do Porto.

Anais de Faculdade de farmacia da
Universidade do Recife. Recife

Anais da Faculdade de medicina da
Universidade do Recife. Recife

Anais da Faculdade de medicina da
Universidade de São Paulo

Anais do Instituto de medicina
tropica. Lisboa

Anais do Instituto superior de
agronomia da Universidade tecnica
de Lisboa. Lisboa

Anais de microbiologia. Rio de
Janeiro

Anais paulistas de medicina e
cirurgia. São Paulo

Anais dos Serviços de veterinária
e indústria animal, Moçambique.
Lourenço Marques.

Anais da Sociedade de biologia
de Pernambuco. Recife.
Anale Institutul de cercetari
agricole. București.

Analele. Institutul Central de
Cercetări Agricole. Secției
de Protecția Plantelor

Analele Institutului de cercetări
agronomice. București.

Analele Institutului de cercetări
pentru cereale si plant tehnice,
Fundulea, Series B, C. Romania

Analele știintifice de
Universitátii "Al. I. Cuza"
din Iași. Section IIa Biologie.
Iași.

Analele Universitátii București,
Seria Stiintele Naturii.
Bologie. București

Anales de bromatologia. Madrid

Anales cientificos. Lima

Anales cientificos. Departamento
de publicaciones de la
Universidad Agraria. La Molina

Anales de edafologia y fisiologia
vegetal. Madrid

Anales de edafologia y agrobiologia
Madrid

Anales de la Escuela nacional de
ciencias biológicas. Mexico

Anales de la Escuela de peritos
agricolas y superior de
agricultura. Barcelona

Anales de la Estación experimental
de Aula Dei. Zaragoza.

Anales de la Facultad de medicina.
Universidad de Montevideo

Anales de la Facultad de medicina.
Universidad nacional mayor de
San Marcos de Lima

Anales de la Facultad de Quimica
y Farmicia, Universidad de
Chile. Santiago

Anales de la Facultad de
veterinaria de León, /Universidad
de Oviedo./

Anales de la Facultad de
veterinaria de la Universidad
de Madrid. Madrid.

Anales de la Facultad de
veterinaria del Uruguay.
Montevideo

Anales del Instituto de biologia.
Universidad de Mexico

Anales del Instituto botanico
A.J. Cavanilles. Madrid

Anales del Instituto forestal de
investigaciones y experiencias.
Madrid.

Anales del Instituto de Higiene
de Montevideo

Anales Instituto de investigaciones
agronomicas. Madrid

Anales del Instituto de
investigaciones veterinarias.
Madrid

Anales del Instituto de medicina
regional. Resistencia.

Anales del Instituto de medicina
regional. Tucumán.

Anales. Instituto nacional de
investigaciones agronómicas.
Madrid

Anales del Instituto nacional
de micro-biologia. Buenos
Aires

Anales del Laboratorio central
S.C.I.S.P. Cochabamba, Bolivia

Anales de lactologia y quimica
agricola de Zaragoza. Zaragoza

Anales de medicina y cirugia.
Barcelona

Anales de medicina pública.
Santa Fé.

Anales. R. Academia de farmacia.
Madrid

Anales de la R. Academia de
medicina. Madrid

Anales de la Sociedad de biologia
de Bogotá.

Anales de la Sociedad cientifica
argentina. Buenos Aires

Analise social. Lisboa.

Analyse et Prévision. Paris

Analyst. London

Analytica chimica acta. New York,
Amsterdam

Analytical Abstracts. Cambridge

Analytical Biochemistry. New York,
London

Analytical Chemistry. Easton, Pa.

Anatomical Record. Philadelphia.

Anatomischer Anzeiger. Jena

Andean airmail and Peruvian times.
Lima

Andhra Agricultural Journal.
Bapatla

Anesthésie et analgésie. Paris

An Foras Taluntais (The Agricultural Institute). Dublin

Angewandte Botanik. Berlin

Angewandte Chemie. Berlin

Angewandte Chemie, International
Edn. Berlin

Angewandte Meteorologie. Berlin

Angewandte Parasitologie. Jena

Angewandte Pflanzensoziologie.
(Institut fur Angewandte
Pflanzensoziologie des Landes
Karnten.) Wien.

Anglo-Swedish Review

Animal Behaviour. London

Animal Breeding Abstracts.
Edinburgh, Farnham Royal

Animal Health Leaflet. Ministry
of Agriculture, Fisheries and
Food. London

Animal Health Yearbook. FAO Rome

Animal Husbandry. Tokyo

Animal Production. Edinburgh

Ankara Üniversitesi tip Fakultesi
Mecmuasi. Ankara

Ankara Üniversitesi Ziraat
Fakültesi yayinlari. Ankara

11

Annalen der Gemeinwirtschaft.
Liege.

Annalen des Naturhistorischen
Museums in Wien. Wien

Annales de l'ACFAS. Association
canadiennefrançaise pour
l'avancement des sciences.
Montreal.

Annales de l'abeille. Paris

Annales Academiae regiae
scientiarum upsaliensis. Sweden

Annales Academiae scientiarum
fennicae.

Series A.II. Chemica
Series A.IV. Biologica

Annales Agriculturae Fenniae.
Helsinki.

Annales agronomiques. Paris

Annales de l'amélioration des
plantes. Paris

Annales de biologie animale,
biochimie et biophysique. Paris

Annales biologicae Universitatum
hungariae. Budapest

Annales bogoriensis. Bogor.
Indonesia.

Annales botanici fennici. Finland

Annales botanici fennici
Societatis zoologicae-botanicae
fennicae Vanamo. Finland

Annales du Centre de Recherches
agronomiques de Bambey au
Sénégal.

Annales de chimie. Paris

Annales chirurgiae et gynaecologiae
fenniae. Helsinki

Annales de dermatologie et de
syphiligraphie. Paris

Annales de la Direction des Etudes
et de l'Equipment. Service
d'Exploitation Industrielle des
Tabacs et des Allumettes.
Section 2. Paris

Annales de l'École nationale
d'agriculture d'Alger. Alger.

Annales de l'École nationale
d'agriculture de Montpellier.

Annales de l'École nationale des
eaux et forêts et de la station
de recherches et expériences
forestières. Nancy

Annales de l'École nationale
supérieure agronomique de
Montpellier

Annales de l'École nationale
supérieure agronomique. Toulouse

Annales d'Endocrinologie. Paris

Annales entomologici fennici.
Finland

Annales des épiphyties et de
phytogénétique. Paris

Annales de la Faculté des sciences
de Marseille. Marseille

Annales des falsifications et de
l'expertise chimique. Paris

Annales de Gembloux. Bruxelles

Annales de génétique. Paris

Annales de géographie. Paris

Annales d'histochimie. Nancy

Annales historico-naturales Musei
nationalis hungarici. Budapest

Annales d'hygiène publique,
industrielle et sociale. Paris

Annales de l'Institut national
agronomique. Paris

Annales de l'Institut national de
la recherche agronomique. Paris

(Series to the above publication
are shown separately)

Annales de l'Institut national de
la Recherche Agronomique. Tunisia

Annales de l'Institut Pasteur.
Paris

Annales de l'Institut Pasteur de
Lille.

Annales de l'Institut phyto-
pathologique Benaki. Athenes.

Annales Instituti biologici,
Tihany, Hungaricae academiae
scientiarum. Tihany.

Annales de médicine. Paris

Annales medicinae experimentalis
et biologiae fenniae. Helsinki

Annales medicinae internae
fenniae. Helsinki

Annales de médicine légale, de
criminologie et de police
scientifique. Paris

Annales de médicine vétérinaire.
Bruxelles

Annales medico-psychologiques.
Paris

Annales du Musée r. de l'Afrique
central. Série in 4to. Zoologie.
Tervuren

Annales du Musée r. du Congo
belge. (Série in 8vo and Série
in 4to Sciences zoologiques)
Tervuren. Belgium

Annales Musei zoologici polonici.
Warszawa

Annales de la nutrition et de
l'alimentation. Paris

Annales d'oculistique. Paris

Annales d'oto-laryngologie. Paris

Annales paediatriae fenniae.
Helsinki

Annales paediatriae fenniae.
Supplements. Helsinki

Annales paediatrici japonici.
Kyoto

Annales de parasitologie humaine
et comparee. Paris

Annales pharmaceutiques françaises
Paris

Annales de physiologie végétale
Bruxelles

Annales de physiologie vegétale.
Paris

Annales de phytopathologie.
Paris

Annales de la recherche forestière
au Maroc. Rabat

Annales des sciences forestieres.
Paris

Annales des sciences naturelles.
Paris

Zoologie et Biologie Animale
Botanique et Biologie Végétale

Annales scientifiques de l'Universite de Besançon (Botanique)

Annales de la Sociétés belges de
medicine tropicale, de Parasitologie et de mycologie

Annales de la Societe entomologique
de France. Paris

Annales de la Société entomologique
de Quebec

Annales de la Société d'horticulture et d'histoire naturelle de
l'Herault. Montpellier

Annales de la Société nationale
de l'horticulture de France.
Paris

Annales de la Société r. zoologique
de Belgique. Bruxelles

Annales des Sociétés belges de
medicine tropicale, de parasitologie et de mycologie humaine
et animale

Annales de speleologie. Paris

Annales de technologie agricole.
Paris

Annales Universitatis Mariae Curie-
Skłodowska. Lublin

Sect.C. Biologia
"   D. Medicina
"   DD.Medicina veterinaria
"   E. Agricultura
"   H. Oeconomia

Annales Universitatis scientiarum
budapestinensis de Rolando
Eötvös nominatae. Budapest

Annales zoologici fennici

Annales zoologici. Polska Akademia
Nauk. Poland

Annales de zootechnie. Paris

Annaler Lantbrukshögskolans.
Uppsala

Annali della Acoademia d'agricoltura di Torino. Torino

Annali. Accademia italiana di
scienze forestali. Firenze.

Annali. Accademia nazionale di
agricoltura. Bologna

Annali di botanica. Roma

Annali di chimica. Roma

Annali della Facoltà di agraria, Università di Bari

Annali della Facoltà di agraria, Universita di Milano. Milano

Annali della Facoltà di agraria, Università degli studi di Perugia

Annali della Facoltà di agraria, Università di Pisa

Annali della Facoltà di agraria di Portici della R. Università di Napoli

Annali della Facoltà di agraria, Università del S. Cuore

Annali della Facolta di medicina veterinaria di Messina. Italy

Annali della Facolta di medicina veterinaria di Torino. Torino

Annali della Facolta di medicina veterinaria. Universita di Pisa

Annali della Facolta di scienze agrarie della Universita degli studi di Napoli. Portici.

Annali della Facolta de scienze agrarie della Universita degli studi di Torino

Annali della Facolta di scienze agrarie della Universita di Palermo

Annali di geofisica. Roma
Annali Idrologici
Annali d'igiene sperimentale. Torino, Roma.

Annali dell'Istituto Carlo Forlanini. Roma.

Annali del'Istituto sperimentale per la zootecnia. Roma.

Annali dell'Istituto superiore di sanita. Italy

Annali italiani di chirurgia. Napoli

Annali di medicina navale e tropicale. Roma

Annali di microbiologia ed enzimologia. Milano

Annali del Museo civico di storia naturale Giacomo Doria. Genova

Annali di radiologia diagnostica. Bologna

Annali della sanità pubblica. Roma

Annali della sperimentazione agraria. Roma

Annali della Stazione chimico-agraria sperimentale di Roma.

Annali. Stazione sperimentale di risicoltura e delle colture irrigue. Vercelli

Annals of Agricultural Science. University of A'in Shams. Cairo

Annals of Allergy. Minneapolis

Annals of Applied Biology. Cambridge.

Annals of Arid Zone. India

Annals of the Association of American Geographers. Minneapolis

Annals of Botany. London

Annals of the Entomological Society of America. Columbus, etc.

Annals of the Entomological Society of Quebec

Annals of Eugenics. London

Annals of Human Genetics. London

Annals of Internal Medicine. Ann Arbor

Annals and Magazine of Natural History. London

Annals of Mathematical Statistics. Ann Arbor, etc.

Annals of Missouri Botanical Gardens. St. Louis

Annals of the Natal Museum. Pietermaritzburg, etc.

Annals of the New York Academy of Sciences. New York

Annals of Otology, Rhinology, and Laryngology. St. Louis

Annals of Public and Cooperative Economy. Liège

Annals of the Rheumatic Diseases. London

Annals of Surgery. London, Philadelphia

Annals of Tropical Medicine and
Parasitology. Liverpool

Annals of Zoology. Agra

Annalos of the Phytopathological
Society of Japan. Tokyo

Annee biologique. Paris

Annotationes zoologicae et
botanicae. Czech.

Annotationes zoologicae japonenses.
Tokyo

Annuaire de la Faculte d'Agricul-
ture et de Sylviculture de
l'Universite de Skopje.
Yugoslavia
  Agriculture
  Sylviculture

Annual Administration Report.
Tea Scientific Section, United
Planters' Association of Southern
India. Madras.

Annual Blueberry Open House.
New Jersey Blueberry Research
Laboratory. Pemberton

Annual Booklet. Association of
Growers of the New Varieties of
Hops. Faversham

Annual Bulletin. International
Commission on Irrigation and
Drainage. New Delhi

Annual Bulletin. International
Dairy Federation. Bruxelles

Annual Departmental Report.
Director of Agriculture, Fisheries
and Forestry. Hong Kong

Annual Number. National Academy
of Sciences, India. Allahabad

Annual Progress Report. Oak Ridge
National Laboratory. Health
Physics Division.

Annual Review of Biochemistry.
Stanford University, Palo Alto

Annual Review of Entomology.
Stanford, Palo Alto

Annual Review. Canterbury
Agricultural College. Lincoln NZ

Annual Review. Lincoln College. NZ

Annual Review of Medicine.
Stanford University. Palo Alto

Annual Review of Microbiology.
Stanford University. Palo Alto

Annual Review of Pharmacology.

Annual Review of Physiology.
Stanford University. Palo Alto

Annual Review of Phytopathology.
Stanford University. Palo Alto

Annual Review of Plant Physiology.
Stanford University. Palo Alto

Annual Review. Rubber Research
Institute of Ceylon. Dartonfield

Annual Ring. University of Toronto.
Toronto.

Annual of the South Devon Herd
Book Society. UK

Annual Work Progress Report on
Crop Improvement Programme of
Rice, Sugarcane, vegetable and
Field Crops. Directorate of
Rural Affairs, Vietnam and
Chinese Technical Mission to
Vietnam on Crop Improvement.
Saigon

Annuario dell'Istituto e museo de
zoologia dell'Università di
Napoli.

Annuario. Stazione sperimentale di
viticoltura e di enologia di
Conegliano. Treviso

Anthropological Quarterly.
Washington

Antibiotic Medicine and Clinical
Therapy. New York. Washington

Antibiotica et chemotherapia.
Basel, New York

Antibiotics Annual. New York

Antibiotics and Chemotherapy.
New York

Antibiotiki. Moskva

Anti-Locust Bulletin. Anti-Locust
Research Centre. London

Anti-Locust Memoir. Anti-Locust
Research Centre. London

Antimicrobial Agents and Chemo-
therapy. Proceedings of the
Interscience Conference on
Antimicrobial Agents and
Chemotherapy. Ann Arbor

Antioquía médica. Medellin

Antiseptic. Edinburgh and Madras

Antonie van Leeuwenhoek. Journal of Microbiology and Serology. Amsterdam

Anuar Institulului de patologie și igiena animala. București

Anuário brasileiro de economia florestal. Rio de Janeiro

Anuário da Sociedade broteriana. Coimbra.

Anzeiger für Schädlingskunde. Berlin, Hamburg

Apicultural Abstracts. London

Appendix to Report. Welsh Soils Discussion Group. Aberystwyth

Applied Microbiology. Baltimore

Applied Science Research Corporation Thailand Newsletter

Applied Statistics. London, Edinburgh

Appraisal Journal. Chicago

Aquarien- und Terrarien-Zeitschrift. Stuttgard

Aquarist and Pondkeeper. London

Arable Farmer. UK

Araneta Journal of Agriculture. Malabou

Arbeit und Socialpolitik, Ausgabe A. Baden-Baden

Arbeiten. Bundesforschungsanstalt für Forst- und Holzwirtschaft. Reinbek

Arbeiten Forschungsgesellschaft für Agrarpolitik und Agrarsoziologie. Bonn

Arbeiten aus dem Gebiete des Futterbaues. Arbeitsgemeinschaft zur Förderung des Futterbaues. Switzerland

Arbeiten aus dem Gebiete des Futterbaues. Experiences fourrageres Articoli di foraggicoltura.

Arbeiten, Institut für Landwirtschaft, Betriebs- und Arbeitsökonomik Gundorf Deutsche Akademie für Landwirtschaft Wissenschaft. Berlin

Arbeiten aus dem Institut für Tierzucht, Vererbungs- und Konstitutionsforschung. Munchen

Arbeiten aus der Medizinischen Fakultät zu Okayama. Okayama

Arbeiten zur Rheinischen Landeskune, Geographischen Institut der Universität Bonn

Arbeitsökonomik. Berlin

Årbog for gartneri. København

Árbók landbunadarins. Iceland

Árbok. Norges Geologiske Undersökelse. Oslo

Arbok for Universitetet i Bergen. Natenatisk-naturvitenskapelig serie. Oslo

Arbor. Aberdeen University Forestry Society. Aberdeen

Arboretum kórnickie. Poznan.

Arboretum Leaves. Ohio

Arboricultural Association Journal. Stansted

Arboriculture fruitière. Paris

Arborist's News. Columbus

Archiv der Deutschen Landwirtschafts-gesellschaft. Hannover.

Archiv für experimentelle Veterinär-medizin. Leipzig, Berlin.

Archiv für experimentelle Pathologie und Pharmakologie. Leipzig

Archiv für Forstwesen. Berlin

Archiv für Gartenbau. Berlin

Archiv für Geflügelkunde. Berlin

Archiv für Geflügelzucht und Kleintierkunde. Berlin

Archiv für die gesamte Virusforschung. Wien

Archiv für Gewerbepathologie und Gewerbehygiene. Berlin

Archiv für Gynaekologie. Berlin

Archiv für Hydrobiologie. Stuttgart

Archiv für Hygiene und Bakteriologie. München u. Berlin.

Archiv der Julius Klaus-Stiftung
für Vererbungsforschung,
Sozialanthropologie und
Rassenhygiene. Zurich.

Archiv für Kinderheilkunde.
Stuttgart

Archiv für klinische Chirurgie
vereinigt mit Deutsche Zeitschrift für Chirurgie. Berlin

Archiv für klinische und
experimentelle Dermatologie.
Wien.

Archiv für Lebensmittelhygiene.
Hannover

Archiv für Meteorologie, Geophysik
und Bioklimatologie. Wien

Archiv für Mikrobiologie. Berlin

Archiv für öffentliche und
Freigemeinnützige Unternehmen

Archiv fur Pflanzenschutz. Berlin

Archiv der Pharmazie und Berichte
der Deutschen pharmazeutischen
Gesellschaft. Berlin

Archiv für Psychiatrie und
Nervenkrankheiten. Berlin, etc.

Archiv für Tierernährung. Berlin

Archiv für Tierernährung und
Tierzucht. Berlin

Archiv für Tierzucht. Berlin

Archiv für Toxikologie. Berlin

Archiva Veterinaria. Roumania

Archives d'anatomie microscopique
et de morphologie expérimentale.
Paris

Archives belges de dermatologie
et de syphiligraphie. Bruxelles

Archives belges de médecine
sociale, hygiène, médicine du
travail et médecine légale.
Bruxelles

Archives of Biochemistry. New York

Archives of Biochemistry and
Biophysics. New York

Archives de biologie. Paris etc.

Archives of Dermatology. New York

Archives of Disease in Childhood.
London

Archives of Environmental Health
Chicago

Archives françaises de pédiatrie.
Paris

Archives of General Psychiatry.
Chicago

Archives d'hygiène. Athènes

Archives de l'Institut botanique
de l'Université de Liége.
Bruxelles, etc.

Archives de l'Institut Razi.

Archives. Institut grand-ducal de
Luxembourg.

  Section des sciences naturelles,
  physiques et mathematiques

Archives de l'Institut Pasteur
d'Algérie. Alger

Archives de l'Institut Pasteur
de la Guyane et de territoire
de l'Inini. Cayenne

Archives de l'Institut Pasteur
hellénique. Athènes.

Archives de l'Institut Pasteur
de Madagascar

Archives de l'Institut Pasteur
de la Martinique. Fort-de-France

Archives de l'Institut Pasteur
de Tunis. Tunis

Archives of Internal Medicine.
Chicago

Archives internationales de
neurologie. Paris

Archives internationales de
pharmacodynamie et de therapie.
Bruxelles, Paris

Archives internationales de
physiologie et de biochimie.
Liege, Paris

Archives internationaux de
Sociologie et co-operation.
Paris

Archives italiennes de biologie.
Pisa

Archives des maladies de
l'appareil digestif et de la
nutrition. Paris

Archives des maladies du coeur,
des vaisseaux et du sang. Paris

Archives des maladies profession-
elles de médecine du travail
et de sécurité sociale. Paris

Archives de médecine générale et
tropicale. Marseille

Archives du Museum national
d'histoire naturelle. Paris

Archives néerlandaises de
zoologie. Leiden

Archives of Neurology. Chicago

Archives d'ophthalmogie. Paris

Archives of Ophthalmology. New York

Archives of Oral Biology. Elmsford

Archives of Otolaryngology. Chicago

Archives of Pathology. Chicago

Archives des recherches agronomiques et pastorales au Viêt Nam.
Saigon

Archives roumaines de pathologie
experimentale et de microbiologie. Paris, Bucureşti

Archives des sciences physiologi-
ques. Paris

Archives of Surgery. Chicago

Archives de zoologie expérimentale
et générale. Paris

Archivio botanico e biogeografico
italiano

Archivio di chirurgia del torace.
Firenze.

Archivio "De Vecchi" per l'anatomia patologica e la medicina
clinica.

Archivio E. Maragliano di patologia e clinica. Genova.

Archivio di fisiologia. Firenze

Archivio italiano di anatomia e
istologia patologica. Milano

Archivio italiano di chirurgia.
Bologna

Archivio italiano dei malattie
dell' apparato digerente.
Bologna

Archivio italiano di otologia,
rinologia e laringologia.
Torino-Palermo

Archivio italiano di pediatria
e puericoltura. Bologna

Archivio italiano di scienze
mediche tropicale e di
parassitologia. Tropoli

Archivio di ortopedia. Milano

Archivio veterinario italiano.
Milano

Archivio zoologico italiano.
Napoli

Archivos argentinos de dermatologia. Buenos Aires

Archivos argentinos de tisiologia.
Buenos Aires

Archivos de biologia. São Paulo

Archivos brasileiros de
cardiologia. São Paulo

Archivos brasileiros de
endocrinologia (e metabologia).
Rio de Janeiro

Archivos brasileiros de medicina.
Rio de Janeiro

Archivos brasileiros de medicina
naval. Rio de Janeiro

Archivos brasileiros de nutricao
Rio de Janeiro

Archivos de cirurgia clinica e
experimental. São Paulo

Archivos de clinica. Rio de Janeiro

Archivos da Escola medico-
cirurgica de (Nova) Goa. Bastora

Archivos da Escola de Veterinaria
da Universidade de Minas Gerais

Archivos españoles de urologia.
Madrid

Archivos da Faculdade de higiene
e Saúde pública da Universidade
de São Paulo

Archivos da Faculdade da medicina Universidade de Bahia

Archivos de Gastroenterologia. São Paulo

Archivos de Higiene e saúde pública. São Paulo

Archivos del Hospital universitario. Habana

Archivos dos hospitals da Santa Casa de São Paulo.

Archivos. Instituto aclimatacion. Almeria.

Archivos do Instituto Bacteriologico Camara Pestana. Portugal

Archivos do Instituto biológico. São Paulo

Archivos do Instituto de pesquisas agronomicas. Pernambuco

Archivos do Instituto de pesquisas veterinarias Desiderio Finamor. Porto Alegre

Archivos internacionales de la hidatidosia. Montevideo

Archivos do Jardim botânico Rio de Janeiro

Archivos de medicina infantil. Habana

Archivos médicos de Cuba. Habana

Archivos médicos municipais. São Paulo

Archivos médicos panamenos. Panama

Archivos médicos de San Loranzo

Archivos do Museu nacional. Rio de Janeiro

Archivos de neuropsiquiatria. São Paulo

Archivos de oftalmologia de Buenos Aires

Archivos de pediatria del Uruguay Montevideo

Archivos do Serviço florestal do Brasil. Rio de Janeiro

Archivos de la Sociedad de biologia de Montevideo

Archivos de la Sociedad oftalmológica hispano-americana. Barcelona

Archivos uruguayos de medicina, cirugia y especialidades. Montevideo

Archivos venezolanos de medicina tropical y parasitologia medica Venezuela

Archivos Latinoamericanos Nutrición. Caracas

Archivos venezolanos de patologia tropical y parasitologia medica. Caracas

Archivos venezolanos de puericultura y pediatria. Caracas

Archivos de zoologia do Estado de São Paulo

Archivos de zootecnia. Córdoba

Archiwum immunologii i terapii doswiad czakbej. Warszawa

Arctic: Journal of the Arctic Institute of N. America. Montreal. New York.

Ardea. Nederlandsche ornithologische vereeniging. Wageningen

Arhiv bioloskih nauka. Beograd

Arhiv za higijenu rada. Zagreb

Arhiv za poljoprivredne nauke (i tehniku). Beograd

Arid Zone Research. UNESCO. Paris

Aristoteleion Panepistemion Thessakolikis Epetiris Tis Geoponikis Kai Dasologikis Skolis. Thessalonikh

Arizona Farmer

Arizona Forestry Notes. School of Forestry, Flagstaff.

Arizona Medicine. Phoenix

Arkansas Farm Research. Fayetteville

Arkhiv patologii. Moskva

Arkiv för botanik. Uppsala. Stockholm

Arkiv för kemi. Stockholm

Arkiv för zoologi. Uppsala

Arnoldia. Arnold Arboretum. Mass.

Arnoldia. Salisbury. Rhod.

ARQUIVOS /incorporated with
ARCHIVOS/

Årsberattelsa. Föreningen för
växtförädling av strogstrad.
Stockholm

Årsberetning. Institut for Steril-
itets-forskning. Denmark

Kongelige Veterinaer- og
Landbohøjskole

Årsberetning, Norsk Treteknisk
Institutt. Blindern

Årsberetning det Norske
Skogforsøksvesen Shogforsknings-
gruppen. Oslo

Årsberetning og regnskap.
Papirindustriens forsknings-
institutt, Vinderen. Oslo

Arsberetning, Skogbrukets og
Skogindustrienes Forskningsforening. Oslo

Arsberetning, Skogbrukets og
Skogindustrienes Forsknings-
forenings Skogforskningsgruppe.
Vollebekk

Årsberetning. Statens skadedyr-
laboratorium. Springforbi.

Årsbok, Finlands Flottareforening.
Helsinki

Årsbok. Föreningen för skogsträds-
förädling. Uppsala

Årsbok. Svenska flottledsförbundet
Stockholm

Årsbok. Sveriges geologiska
undersokning. Stockholm

Arsmelding, Institutt for
Skogøkonomi, Norges Landbruk-
shøgskole. Vollbekk

Arsmelding. Norges landbrukssviten-
skapelige forskningsrad. Oslo

Årsmelding. Skogdirectørin. Oslo

Årsrit Raektunarfelage Nordurlands.
Iceland

Årsrit. Shograektarfelage Islands
Reykjavik

Årsskrift. Kalmar Läns Södra
Hushallningssallskap

Årsskrift. Norske Skogplanteskoler
Oslo

Årsskrift. K. Veterinaer- og
Landbohøjskole. København

Artha Vijnana. Poona

Artha Vikas. Vallabh Bidyanagar

Arzneimittel-Forschung. Aulendorf

Aquatic Biology Abstracts.London

Asian Economic Review. Hyderabad

Asian Review. London

Asian Survey. Berkely. Calif.

Association of Agriculture Journal
London

Association Forêt-Cellulose
(AFOCEL) Paris

Atas. Instituto de micologia
Universidade do Recife.
Atas da Sociedade de biologia do
Rio de Janeiro

Ateneo parmense. Parma

Athena: rassegna mensile di
biologia, clinica e terapia.
Roma

Atlantic Community Quarterly.
Baltimore

Atlantide Report. Scientific Results
of the Danish Expedition to the
Costs of Tropical West Africa.
1945-46. Copenhagen

Atoll Research Bulletin.Washington

Atomic Energy Review. Vienna

Atomnaya Energiya. Moskva

Atompraxis. Karlsruhe

Atti. Accademia italiana della
vite e del vino siena.
Firenze

Atti dell' Accademia nazionale dei
Lincei. Rendiconti. Roma

Atti dell' Accademia delle
scienze. Torino

Atti dell' Accademia delle
Scienze dell' Istituto di
Bologna. Memorie Ser.II Classe
di Scienze Fisiche (Bologna)

Atti dell' Accademia delle
Scienze dell' Istituto di
Bologna. Rendiconti. Ser.XII
Classe di Scienze Fisiche
(Bologna)

Atti dell' Instituto botanico
della Universita e Laboratorio
crittogamico di Pavia. Milano

Atti. Associazione genetica
italiana Roma.

Atti e memorie dell Accademia
d'agricoltura, scienze e
lettere di Verona. Verona

Atti della Società italiana di
scienze naturali, e del Museo
civile di storia naturale.
Milano

Atti della Societa italiana delle
scienze veterinarie. Faenza

Atti del V. Simposio Internazionale di Agrochimica su "Lo Zolfo
in Agricultura". Palermo

Auburn Veterinarian. Auburn. Ala

Augsne un raža. Riga

Auk. A quarterly journal of
ornithology. Cambridge. Mass.

Aus Politik und Zeitgeschehen.
Bonn

Aus dem Walde. Hannover

Aussenpolitik. Freiburg

Australasian Annals of Medicine.
Sydney

Australasian Irrigator and Pasture

Australian Citrus News

Australian Forestry. Canberra

Australian Forest Research.
Canberra

Australian Geographer. Sydney

Australian Grape Grower.

Australian Journal of Agricultural
Economics. Sydney

Australian Journal of Agricultural
Research. Melbourne

Australian Journal of Applied
Science. Melbourne

Australian Journal of Biological
Sciences. Melbourne

Australian Journal of Botany.
Melbourne

Australian Journal of Chemistry.
Melbourne

Australian Journal of Dairy
Technology. Melbourne

Australian Journal of Dermatology
Sydney

Australian Journal of Experimental
Agriculture and Animal
Husbandry. Melbourne

Australian Journal of Experimental
Biology and Medical Science.
Adelaide

Australian Journal of Instrument
Technology. Melbourne.

Australian Journal of Marine and
Freshwater Research. Melbourne

Australian Journal of Science
Sydney

Australian Journal of Soil
Research

Australian Journal of Zoology
Melbourne

Australian Museum Magazine.
Sydney

Australian and New Zealand Journal
of Surgery. Sydney

Australian Oil Seed Grower

Australian Paediatric Journal.
Melbourne

Australian Plant Disease Recorder.
Sydney

Australian Primary Producers'
Union Review. Melbourne

Australian Quarterly. Sydney

Australian Science Index.
Melbourne

Australian Sugar Journal.
Brisbane

Australian Timber Journal.
Sydney

Australian Tobacco Growers'
Bulletin

Australian Tobacco Journal

Australian Veterinary Journal

Autogestion. Paris

Automation in Analytical Chemistry Technicon Symposia. Paris

Avances en Alimentacion y Mejora Animal. Madrid

Avian Diseases. Ithaca. NY.

Al-Awamia. Review de la Recherche Agronomique Marocaine. Institut National de la Recherche Agronomique. Rabat.

Ayrshire Cattle Society's Journal. Ayr.

B

B.A.S.F. bildbericht vom Limburgerhof

B.C.F.G.A. Quarterly Report. British Columbia Fruit Growers' Association. Kelowna.

B.G.R. Business and Government Review, Columbia. Mo.

B.R.G.M. Bulletin. Paris

B.S.I. News. British Standards Institution. London

B.W.P.A. News Sheet. British Wood Preserving Association. London

Bacteriological Proceedings. Society of American Bacteriologists. Baltimore

Bacteriological Reviews. Baltimore

Baileya. Quarterly journal of horticultural taxonomy. New York

Bainne. Dublin

Balwant Vidyapeeth Journal of Agricultural and Scientific Research

Bamidgeh. Bulletin of Fish Culture in Israel. Nir-David

Banana Bulletin. Murwillumbah. N.S.W.

Bangkok Bank Monthly Review. Bangkok

Bank of London and South America Review. London

Bankers' Magazine. London

Banque Centrale des Etats de l'Afrique de l'Ouest. Notes d'information et Statistiques. Paris

Banque française et italienne pour l'Amerique du sud: note mensuel. Paris

Barclay's Bank Review. London

Baromfipar. Budapest

Baromfitenyésztés. Budapest

Basteria: Tijdschrift van de Nederlandsche malacologische vereeniging. Lisse

Bayerisches landwirtschaftliches Jahrbuch. Munchen

Baywood Courier. London

Beaufortia. Miscellaneous Publications of the Zoological Museum, Amsterdam

Bee World. Oxford. etc.

Beef Production Handbook. Beef Recording Association (UK) Reading

Beef Research Report. Bureau of Agricultural Economics. Canberra

Beef and Sheep Farming. UK

Beef Shorthorn Record. UK

Beet Grower. Dublin

Behaviour: an international journal of comparative ethnology. Leiden

Behavioural Sciences and Community Development. Hyderabad

Beihefte zur Nova Hedwigia. Weinheim

Beiträge zur Biologie der Pflanzen. Breslau.

Beiträge zur Entomologie. Berlin

Beiträge zur Klinik der Tuberkulose und zur spezifischen Tuberkuloseforschung. Würzburg.

Beiträge zur klinischen chirurgie. Tübingen

Beiträge zur Mineralogie und
Petrographie. Berlin, etc.

Beiträge zur Naturkunde
Niedersachsens. Osnabrück.

Beiträge zur pathologischen
Anatomie und zur allgemeinen
Pathologie. Jena

Beiträge zur tropischen und sub-
tropischen Landwirtschaft und
tropenveterinärmedizin.
Germany

Belgique laitière. Brussels

Belgisch tijdschrift voor
geneeskunde. Gent.

Bengal Veterinarian.

Beplantingen en Boomkweekerij

Beretning om Faellesforsøg i
Landboog Husmandsforeningerne.
Denmark

Beretning fra Forsøgslaboratoriet.
København

Beretning om Landboforeningernes
Virksomhed for Planteavlen paa
Sjaelland

Beretning om Planteavien paa
Lolland-Falster. Nykøbing

Beretning fra Statens Forsøgs-
mejeri. Hillerød

Beretning fra Statsfrøkontrollen.
Denmark

Beretning fra Statens forsøksvirksomhed i plantekultur.
Copenhagen

Beretning fra Statens Planteavls-
udvalg. København

Bergcultures. Djakarta

Berichte Arbeitsgemeinschaft zur
Verbesserung der Agrarstruktur
in Hessen. Wiesbaden

Berichte über die Arbeitstagung
der Arbeitsgemeinschaft der
Saatzuchtleiter. Gumpenstein.

Bericht der Bayerischen
botanischen Gesellschaft zur
Erforschung der heimischen
Flora. München

Bericht der Deutsche Gesellschaft
für Holzforschung. Stuttgart

Bericht der Deutschen botanischen
Gesellschaft. Berlin

Bericht des Deutschen
Wetterdienstes. Offenbach

Bericht des Geobotanischen
Institutes der Eidg. Techn.
Hochschule, Stiftung Rübel.
Switzerland

Bericht und informationen.
Salzburg.

Bericht des Instituts für
Tabakforschung. Dresden

Berichte aus der Land- und
Forstwirtschaftlichen Forschung.
Germany

Bericht aus der Landesanstalt für
Bodennutzungsschutz des Landes
Nordrhein-Westfalen. Bochum.

Bericht über Landwirtschaft.
Berlin

Bericht der Naturforschenden
Gesellschaft zu Freiburg i Br.

Bericht der Oberhessischen
Gesellschaft für Natur- u.
Heilkunde. Giessen

Bericht des Ōhara Instituts für
landwirtschaftliche Biologie.
Kuraschiki.

Bericht over rassenkeuze.
's Gravenhage

Bericht der Schweizerischen
botanischen Gesellschaft. Bern

Bericht. Stichting Bosbouwproefstation 'De Dorschkamp'.

Berichte über die Tagung.
Nordwestdeutscher Forstverein.
Hannover

Bericht und Vorträge der Deutschen
Akademie der Landwirtschafts-
wissenschaften zu Berlin.
Berlin

Berita^2 dari Perusahaan^2 Gula di
Indonesia.

Berliner und Münchener tierärztliche Wochenschrift. Berlin

Berufsdermatosen. Aulendorf.

Besseres Obst.

Beten, vallar, mossar. Uppsala

Better Crops with Plant Food. New York

Better Fruit. Portland, Ore.

Better Plants Better Packs.

Biatas. Dublin

Bibliographia genetica. s' Gravenhage.

Bibliographical Series. Oregon State University Forest Research Laboratory. Corvallis

Bibliographie agricole courante roumaine

Bibliographie der Pflanzenschutz-literatur. Berlin

Bibliographien des deutschen Wetterdienstes. Bad Kissingen

Bibliography of Agriculture. US Department of Agriculture. Washington

Bibliography (Annotated). Commonwealth Forestry Bureau. Oxford.

Bibliography. Commonwealth Bureau of Soils. Harpenden. Herts.

Bibliography, Documentation and Terminology. UNESCO

Bibliography on Pesticide Toxicity and Accidental Poisoning. WHO. Geneva

Bibliography of Reproduction. UK

Bibliography of Scientific Publications of South and South East Asia. New Delhi

Bibliography of Systematic Mycology. Kew

Bibliotheca Haematologica. Basle

Bibliotheca 'Nutritio et dieta' Basel and New York

Bibliotheca paediatrica. Basel

Biennial Report. Division of Plant Industry. Department of Agriculture. Florida.

Biennial Report. Hawaii University Agricultural Experiment Station. Honolulu.

Biennial Report. Kansas Agricultural Experiment Station, Manhattan. Topeka.

Bihar Animal Husbandry News.

Bijdragen tot de dierkunde. Leiden.

Biken's Journal. Osaka

Bilag til Arsberetning fra Statens Forsøgmejeri. Hillerød.

Bi-monthly Research Notes. Forestry Branch. Department of Forestry and Rural Development. Ottawa

Bi-monthly Bulletin. North Dakota Agricultural Experiment Station. Fargo.

Biochemical and Biophysical Research Communications. New York, London.

Biochemical Genetics. New York

Biochemical Journal. Liverpool, Cambridge, London

Biochemical Medicine. New York

Biochemical Pharmacology. London, New York

Biochemical Preparations. New York

Biochemical Society Symposia. Cambridge. UK

Biochemistry. New York

Biochemistry of Sulfur Isotopes. Proceedings of a National Science Foundation Symposium, Yale University, 1962. New Haven

Biochimica et biophysica acta. New York, Amsterdam

Biodynamica. Normandy, Mo.

Biofizika. Moskva.

Biographical Memoirs of Fellows of the Royal Society. London

Biokhimiya. Leningrad, Moskva.

Biokemia. Midland, Mich.

Biologia. Biological Society of Pakistan. Lahore

Biológia. Casopis Slovenskej
akademie vied. Bratislava.

Biologia neonatorum. Basel

Biologia plantarum. Praha

Biologica. Trabajos del
Instituto de biologia de la
Universidad de Chile. Santiago

Biologica latina. Milano

Biological Abstracts. Menasha.
Philadelphia

Biological Bulletin. Marine
Biological Laboratory, Woods Hole
Mass.

Biological Journal. Linnean
Society

Biological Journal of Okayama
University. Okayama

Biological Reviews and Biological
Proceedings of the Cambridge
Philosophical Society.

Biological Studies. Catholic
University of America. Washington

Biologicheskie nauki. Erevan

Biologicheskii Zhurnal Armenii.
USSR

Biologické pracé Slovenskej
akademie vied. Bratislava

Biológico. São Paulo

Biologie medicale. Paris

Biologie du Sol. Bulletin
International d'Informations.
(Association Internationale de
la Science du Sol.) Paris

Biologisch jaarboek. Antwerpen

Biologische Beiträge. Germany

Biologische Rundschau

Biologisches Zentralblatt.
Leipzig

Biologizace a Chemizace Výživy
Zvirat. Prague

Biology and Human Affairs. London

Biološki glasnik. Zagreb

Biološki Institut N.R. Srbye
Posebna izdanja izdanja

Bio-Medical Engineering. London

Biometrics. American Statistical
Association. Washington

Biometrics Bulletin. American
Statistical Association.
Washington

Biométrie-praximétrie. Brussels

Biometrika. Cambridge, London

Biometrische Zeitschrift. Berlin

Biophysical Journal. New York

Biopolymers. New York

BioScience. USA

Biota. Instituto salesiano 'Pablo
Albera'. Magdalena del Mar

Biotechnology and Bioengineering.
USA

Bird Study. Journal of the
British Trust for Ornithology.
London, Oxford.

Bitki koruma bülteni. Ankara.

Biuletyn, Instytut Badawczy
Lesnictwa. Warsaw

Biuletyn Instytutu Hodowli i
Aklimatyzacji Roślin. Warsaw

Biuletyn Instytutu Ochrony Roślin.
Poznań

Biuletyn Instytutu roślin
leczniczych. Poznań

Biuletyn, Instytut Technologii
Drewna. Poznań

Biuletyn peryglacjalny. Łódź

Biuletyn Slaski Instytut Maukowy.
Katoweś

Biuletyn warzywniczy. Poland

Biuletyn. Zakład Hodowli
Doswiadczalnej Zwierząt Polskiej
Akademii Nauk. Poland

Biztositasi Szemle. Budapest

Blauen Hefte für den Tierarzt.
Germany

Blood. The Journal of hematology.
New York

Blue Book for the Veterinary
Profession. Germany

Blumea. Tijdschrift voor de
systematiek en de geografie der
planten. Leiden

Blut. Zeitschrift für
Blutforschung. Germany

Blyttia. Norsk botanisk forenings
tidsskrift. Oslo

Board of Trade Business Monitor
Production Series. Pesticides
and Allied Products. London

Bodenkultur. Wien

Bois et forêts des tropiques.
Paris

Boletim do Centro de estudos do
Hospital dos servidores do
estado. Rio de Janeiro

Boletim clinico dos Hospitais civis
de Lisboa

Boletim. Commissão reguladora das
cereais do arquipelaga. Azores

Boletim cultural da Guiné
portuguesa. Lisboa

Boletim da Directoria da producção
animal, Rio Grande do Sul.
Pôrto Alegre

Boletim fitossanitario. Divisão
de defesa sanitária vegetal.
Rio de Janeiro

Boletim de Fundação Gonçalo
Moniz. Bahia.

Boletim do Hospital das clinicas
da Faculdade de medicina da
Universidade da Bahia. Salvador

Boletim de industria animal.
São Paulo

Boletim do Instituto biologico da
Bahia. São Salvador

Boletim do Instituto oceanográfico
São Paulo

Boletim Junta Nacional do Azeite.
Lisboa

Boletim Junta Nacional da cortiça.
Lisboa

Boletim. Hunta Nacional das frutas.
Lisboa

Boletim do leite e seus derivados.
Rio de Janeiro

Boletim do Museu de biologia Prof.
Mello-Leitão. Santa Teresa.

Boletim do Museu nacional de
Rio de Janeiro.

Nova série. Zoologia

Boletim do Museu paraense
'Emilio Goeldi'. Pará.
Ser. Zoologia.

Boletim da Ordem dos engenheiros.
Lisboa

Boletim paranaense de geografia.
Curitiba

Boletim pecuário. Lisboa

Boletim dos servicos de saúde
pública. Lisboa

Boletim. Setor de Inventarios
Florestais, Seçao de Pesquisas
Florestais, Divisao de Silvi-
cultura, Rio de Janeiro

Boletim da Sociedade Broteriana.
Coimbra.

Boletim da Sociedade de estudios
(da Colónia) de Moçambique.
Lourenço Marques.

Boletim da Sociedade paulista de
medicina veterinaria. São Paulo

Boletim da Sociedade portuguesa de
ciências naturais. Lisboa

Boletim da Superintendencia dos
serviços do café. São Paulo

Boletim técnico. Instituto
agronômico del Norte. Belem.

Boletim técnico. Instituto
agronômico do sul. Pelotas,
Pôrto Alegre.

Boletim técnico do Instituto de
Pesquisas e Experimentação
agropecuarias do Norte. Belem

Boletin agro-pecuario. Barcelona

Boletin agrentino forestal.
Buenos Aires

Boletin de la Asociación médica
de Puerto Rico. San Juan, etc.

Boletin de la Camara Oficial
Sindical Agraria de Barcelona

Boletin del Centro de
Investigaciones Veterinarias
Miguel C. Rubino. Uruguay

Boletin chileno de parasitologia.
Santiago de Chile

Boletin de la Compañia
administradora del guano. Lima

Boletin de la Dirección general
impositiva. Buenos Aires

Boletin de divulgación ganaderia.
Ciudad Real

Boletin divulgativo. Instituto
Nacional de Investigaciones
Forestales, Secretaria de
Agricultura y Ganaderia, Mexico

Boletin economico de america
latina, Santiago de Chile

Boletin eipidemiológico. Mexico

Boletin. Estación experimental
agronómica de Cuba. Santiago de
las Vegas

Boletin. Estación experimental
agricola de La Molina. Lima

Boletin de la Estación experimental
de Paysandu. Paysandu

Boletin. Estación experimental
agricola de Tucumán. Tucumán.

Boletin Estación Experimental
agricola Universidad de Puerto
Rico. Rio Piedras

Boletin de estudios economicos.
Bilbao

Boletin. Facultad de agronomia,
Universidad de la Republica,
Montevideo

Boletin. Facultad de agronomia y
veterinaria de la Universidad
de Buenos Aires

Boletin Forestal y de Industrias
Forestales para America Latina,
Oficina Forestal Regional de la
FAO Santiago

Boletin Genetico, Instituto de
Fitotecnia. Casteldr.

Boletin Indigenista Venezolana.
Caracas

Boletin de información. Consejo
general de colegios
veterinarios de España. Madrid

Boletin de información del
Ministerio de agricultura
de industria y agricultura.
Madrid

Boletin de informacion tecnica,
AITIM. Madrid

Boletin de informaciones
cientificas nacionales. Quito

Boletin de informaciones
parasitarias chilenas. Santiago

Boletin informativo. Centro
nacional de investigaciones de
café, Colombia. Chinchina

Boletin informativo. Estacion
experimental de Paysandu.
Uruguay

Boletin informativo. Instituto
de fitotecnia. Castelar

Boletin informativo. Instituto
Forestal Santiago

Boletin informativo. Instituto
Salv. investigacion de cafe

Boletin informativo. Instituto
Nacional de Investigaciones
Forestales, Secretaria de
Agricultura y Ganaderia, Mexico

Boletin Informativo. Ministerio
de Ganaderia y Agricultura,
Uruguay

Boletin Informativo Trimestral
Centro Panamericano de Zoonosis.
Argentina

Boletin Informaturo, Junta
Nacional de granas.Buenos Aires

Boletin del Instituto de ciencias
naturales, Universidad central
del Ecuador. Quito

Boletin del Instituto de clinica
quirurgica. Buenos Aires

Boletin. Instituto forestal de
investigaciones y experiencias.
Madrid

Boletin. Instituto forestal
latinoamericano de investigación
y capacitación. Merida

Boletin del Instituto nacional
de investigaciones agronómicas.
Madrid

Boletin del Instituto oceano-
grafico. Universidad de Oriente.
Venezuela

Boletin del Instituto de
investigaciones veterinarias.
Caracas

Boletin del Laboratorio de la
Clinica 'Luis Razetti'.
Caracas

Boletin de Laboratorio Clinico.
Caracas

Boletin médico del Hospital
infantil. Mexico

Boletin mensual. Dirección de
ganaderia. Montevideo

Boletin del Museo argentino de
ciencias naturales 'Bernardino
Rivadavia' e Instituto nacional
de investigación de ciencias
naturales. Buenos Aires

Boletin del Museo de ciencias
naturales. Caracas

Boletin de Noticias. Instituto de
Formento Algodonero. Departmento
de Experimentación. Bogota

Boletin de la Oficina sanitaria
panamericana. Washington

Boletin de oleicultura
internacional. Madrid

Boletin de patologia vegetal y
entomologia agricola. Madrid

Boletin de producción animal.
Chile

Boletin de la R. Sociedad española
de historia natural. Madrid
Seccion Biologica
Boletin de sanidad militar.
Mexico

Boletin de la Sociedad argentina
de botanica. La Plata

Boletin de la Sociedad de biologia
de Concepción (Chile).

Boletin. Sociedad botanica de
Mexico

Boletin de la Sociedad de cirugia
del Uruguay. Montevideo

Boletin de la Sociedad cubana de
dermatologia y sifilografia.
Habana

Boletin de la Sociedad venezolana
de ciencias naturales. Caracas

Boletin del Servicio de plagas
forestales. Madrid

Boletin de la Sociedad quimica
del Peru. Lima

Boletin tecnico. Escuela de
Ingenieria Forestal, Universidad
de Chile, Santiago

Boletin técnico. Facultad de
ciencias agrarias, Universidad
national de Cuyo. Mendoza

Boletin técnico. Federación
nacional de cafeteros de
Colombia. Chinchina

Boletin tecnico. Instituto
Forestal, Santiago

Boletin Técnico. Instituto
Nacional de Investigaciones
Forestales, Secretaria de
Agricultura y Ganaderia.
Mexico

Boletin Técnico. Ministerio de
Agricultura, Guatemala

Boletin Técnico. Ministerio de
Agricultura, Lima. Peru

Boletin y trabajos. Academia
argentina di cirugia.
Buenos Aires

Boletin y trabajos. Sociedad
argentina de cirujanos.
Buenos Aires

Boletin trimestral de
experimentacion agropecuaria.
Lima

Programa cooperativo de experimentacion agropecuaria

Boletin. Universidad de la
Republica. Facultad de
Agronomia. Montevideo

Boletín. Venezolano del
laboratorio clinico. Caracas

Boletin de Zootecnia. Spain

Bollettino di chimica clinica e
farmacoterapia. Napoli

Bollettino dell'Instituto di
entomologia agraria e dell'
Osservatorio di fitopathologia
di Palermo. Palermo

Bollettino dell'Instituto di
entomologia dell Università
degli studi di Bologna

Bollettino dell'Instituto e
Museo di zoologia della
Università di Torino

Bollettino dell'Istituto di patologia del libro. Roma

Bollettino dell'Istituto sieroterapico milanese. Milano

Bollettino dei Laboratori chemici provinciali. Bologna

Bollettino del Laboratorio di entomologia agraria 'Filippo Silvestri' Portici.

Bollettino del Laboratorio sperimentale e Osservatorio di fitopatologia. Torino

Bollettino del Museo civico di storia naturale di Venezia. Venezia

Bollettino schermografico. Roma

Bollettino della Società entomologica italiana. Firenza

Bollettino della Società italiana di biologia sperimentale. Napoli, etc.

Bollettino della Società medico-chirurgica di Modena

Bollettino della Società di naturalisti i Napoli. Napoli

Bollettino della Stazione di patologia vegetale di Roma. Firenze

Bollettino di zoologia, pubblicato dall'Unione zoologica italiana. Napoli

Bollettino di zoologia agraria e bachicoltura. Torino, etc.

Bombay Veterinary College Magazine Bombay

Bondevennen

Bonner geographische Abhandlungen. Bonn

Bonner zoologische Beiträge. Bonn

Bonplandia. Corrientes

Booklet. Forestry Commission. London

Booklet. National Agricultural Advisory Service, Ministry of Agriculture, Fisheries and Food. London

Bosbou in Suid-Afrika. Pretoria

Botanica marina. International Review for Seaweed Research and Utilization. Hamburg

Botanical Bulletin of Academia sinica. Taipei

Botanical Gazette. Chicago

Botanical Magazine. Tokyo

Botanical Museum Leaflets. Harvard University, Cambridge, Mass.

Botanical Journal. Linnean Society

Botanical Review. Lancaster, Pa.

Botanicheskie materialy Gerbariya Botaniki Akademii Nauk Kazakhskoĭ SSR.

Notulae systematicae ex Herbario Instituti Botanici Academiae Scientiarum Kasachstanicae. Alma Ata

Botanicheskiĭ zhurnal SSSR. Leningrad, Moskva

Botanichnii zhurnal. Kyyiv

Botanikai közlemények. Budapest

Botanische Jahrbücher für Systematik, Pflanzengeschichte und Pflanzengeographie. Leipzig, etc.

Botanisk Tidsskrift. Kjøbenhavn

Botaniska notiser. Lund

Bothalia. National Herbarium, Pretoria

Botyu-kagaku. Scientific Insect Control Bulletin of the Institute of Insect Control, Kyoto University. Kyoto

Bragantia. Campinas.

Brain. A Journal of Neurology, London

Brain Research. Amsterdam

Brasil açucareiro. Rio de Janeiro

Bratislavské lékařske listy. Praha

Brazil-medico. Rio de Janeiro

Brewers' Guild Journal. London

British Agricultural Bulletin. London

British Birds. London

British Cattle Breeders' Club Digest. Uckfield, Sussex

British Columbia Lumberman. Vancouver

British Dental Journal. London

British Farmer. London

British Food Journal. London

British Friesian Journal. Lewes

British Heart Journal. London

British Journal of Applied Physics. London

British Journal of Cancer. London

British Journal of Clinical Practice. London

British Journal of Dermatology. London

British Journal of Diseases of the Chest. London

British Journal of Experimental Biology. Edinburgh

British Journal of Experimental Pathology. London

British Journal of Haematology. Oxford

British Journal of Industrial Medicine. London

British Journal of Medical Psychology. London

British Journal of Nutrition. Cambridge

British Journal of Phythalmology. London

British Journal of Pharmacology and Chemotherapy. London

British Journal of Preventative and Social Medicine. London

British Journal of Psychiatry. London

British Journal of Radiology. London

British Journal of Sociology. London

British Journal of Surgery. Bristol

British Journal of Tuberculosis and Diseases of the Chest

British Journal of Urology

British Journal of Venereal Diseases. London

British Medical Bulletin. London

British Medical Journal. London

British National Bibliography

British Poultry Science. Edinburgh, London

British Standard. London

British Standard Specification. London

British Standards Yearbook. London

British Sugar Beet Review. London

British Veterinary Journal. London

Brittonia. New York

Brompton Hospital Reports. London

Broteria. Lisboa

Serie trimestral: Ciencias naturais

Bruinsma Bulletin. Naaldwijk

Bruxelles médical. Bruxelles

Building Science Abstracts. London

Buletin ştiinţific. Academia republicii populare române. Bucureşti

Seria: Ştiinţe medicale

Buletini i Shkencave Bujqësore. Albania

Buletinul Institului Politehnic din Braşov Seria B. Economie Forestieră

Buletinul Institutului politehnic din Iaşi. Iaşi.

Búlgarski tyutyun. Sofia

Bulletin de l'Académie Malgache. Tananarive

Bulletin de l'Académie nationale de medecine. Paris

Bulletin de l'Académie polonaise des sciences. Poland

Serie des Sciences Biologiques

Bulletin de l'Académie r. de Belgique. Classe des sciences. Bruxelles

Bulletin de l'Académie r. de medecine de Belgique. Bruxelles

Bulletin de l'Académie r. des sciences coloniales. Belgium

Bulletin de l'Académie serbe des sciences. Classe des sciences médicales. Beograd

Bulletin de l'Académie suisse des sciences medicales. Switzerland

Bulletin de l'Académie vétérinaire de France. Paris

Bulletin. Afdelingen voor Sociale Wetenschappen aan de Landbouw-hogeschool. Wageningen

Bulletin de l'Afrique noire. Paris

Bulletin agricole

Bulletin agricole du Congo belge Bruxelles, etc.

Bulletin. Agricultural Research Oklahoma State University. Stillwater

Bulletin. Agricultural Research Service, US Department of Agriculture. Washington

Bulletin agronomique. Inst. Rech. agron. trop. cult.

Bulletin. Alabama Agricultural Experiment Station. Auburn.

Bulletin. Alaska Agricultural Experiment Station. Washington

Bulletin. American Iris Society

Bulletin et annales de la Société r. entomologique de Belgique. Bruxelles

Bulletin of the Aomori Agricultural Experiment Station

Bulletin of the Arizona Agricult-ural Experiment Station. Tucson

Bulletin. Arkansas Agricultural Experiment Station. Fayetteville

Bulletin de l'Association des diplomés de microbiologie de la Faculté de pharmacie de Nancy. Nancy

Bulletin de l'Association Française pour l'Étude du Cancer

Bulletin de l'Association Française pour l'Étude du Sol. Paris

Bulletin de l'Association de Geographes Française. Paris

Bulletin of the Association for Tropical Biology. Trinidad

Bulletin of the Azabu Veterinary College. Kanagawa-Ken

Bulletin of the Barbados Department of Agriculture

Bulletin. Barley Improvement (Research) Institute. Winnipeg

Bulletin Bibliographique. Isotopes Rayonnements Agriculture Saint-Paul-lez-Durance (B.-du-R).

Bulletin Bibliographique de Pédologie. Office de la Recherche Scientifique et Technique, Outre-Mer. Paris.

Bulletin of the Biogeographical Society of Japan. Tokyo

Bulletin biologique de la France et de la Belgique. Paris

Bulletin of the Botanical Society of Bengal. Calcutta

Bulletin of the Botanical Survey of India

Bulletin of Brewing Science. Tokyo

Bulletin of the British Museum (Natural History), Entomology. London

Bulletin of the British Mycological Society. Higham

Bulletin of the British Ornithologists' Club. London

Bulletin. British West Indian Central Sugar Cane Breeding Station. Barbados

Bulletin of the Brooklyn Entomological Society. Lancaster, Pa.

Bulletin. Bureau of Biological Research, Rutgers University. New Brunswick

Bulletin of the Calcutta School
of Tropical Medicine (and
Hygiene). Calcutta

Bulletin of the California
Agricultural Experiment Station.
Berkeley

Bulletin of the California
Department of Agriculture.
Sacramento

Bulletin of the California Insect
Survey. Berkeley, Los Angeles

Bulletin. Canada Department of
Agriculture. Ottawa

Bulletin of the Canterbury
Chamber of Commerce.

Bulletin of the Central Research
Institute, University of Kerala.
Trivandrum.

　Series C. Natural sciences

Bulletin des Centres d'études
techniques agricoles. Paris

Bulletin of the Chugoku-Shikoku
Agricultural Experimental
Station. Himeji
Series B.

Bulletin of the Citrus Experiment
Station, Shizuoka Prefecture.
Shimizu-shi

Bulletin of the Clinical and
Scientific Society, Cairo
University. Abbassia

Bulletin. Coconut Research
Institute, Ceylon. Colombo

Bulletin. College of Agriculture,
Forestry and Wood Technology
Laboratory University of Teheran.
Karaj, Persia

Bulletin of the College of
Agriculture and Veterinary
Science, Nihon University

Bulletin of the College of
Agriculture, Utsonomiya
University

Bulletin of the College of Arts
and Sciences. Baghdad

Bulletin. Colorado State
University Agricultural
Experiment Station. Fort Collins

Bulletin. Colorado Flower Growers'
Association

Bulletin du Comité des forêts.
Paris

Bulletin. Commonwealth Bureau of
Pastures and Field Crops.
Aberystwyth

Bulletin. Commonwealth Scientific
and Industrial Research
Organization. Melbourne

Bulletin de Conjonçture régionale.
Rennes

Bulletin. Connecticut Agricultural
Experiment Station. New Haven.

Bulletin of the Connecticut
(Storrs) Experiment Station

Bulletin. Cooperative Extension
Service Montana State
University. Bozeman

Bulletin. Cornell University
Agricultural Experiment Station.
Ithaca, N.Y.

Bulletin. Council for Scientific
and Industrial Research.
Melbourne

Bulletin of current references
on agriculture in India.
New Delhi

Bulletin of the Delaware
University Agricultural
Experiment Station. Newark

Bulletin. Department of Agricult-
ural Economics. North Dakota
State University, Agricultural
Applied Science, Fargo. N.Dakota

Bulletin. Department of
Agricultural Economics,
University of Illinois. Urbana

Bulletin. Department of
Agricultural Economics,
University of Manchester

Bulletin. Department of Agriculture
Bermuda. Hamilton

Bulletin. Department of Agriculture
Fiji. Suva

Bulletin of the Department of
Agriculture, Jamaica

Bulletin. Department of
Agriculture, Federation of
Malaya, Kuala Lumpur

Bulletin. Department of
Agriculture, Mauritius. Reduit

Bulletin. Department of
Agriculture, New Zealand.
Wellington

Bulletin. Department of
Agriculture, Western Australia.
Perth

Bulletin of the Department of
Agriculture and Forestry,
Union of South Africa. Pretoria

Bulletin of the Department of
Agriculture and Lands, Southern
Rhodesia. Salisbury

Bulletin. Department of
Agricultural Technical Services.
Pretoria

Bulletin of the Department of
Forestry, Pretoria

Bulletin. Department of Forestry
University of Ibadan

Bulletin. Department of Resources
and Economics, University of
New Hampshire. Durham.

Bulletin. Department of Rural
Sociology, Cornell University.
Ithaca

Bulletin. Department of Science
and Agriculture, Barbados

Bulletin of the Division of
Agriculture, Federation of Malaya

Bulletin. Division of Plant
Industry, Department of
Agriculture, Florida

Bulletin of the Division of Plant
Industry, Queensland Department
of Agriculture and Stock.
Brisbane

Bulletin. Division of Silviculture.
Department of Forests, Papua and
New Guinea

Bulletin de Documentation

Bulletin de Documentation.
Association internationale des
fabricants de superphosphates.
Paris

Bulletin. Duke University, School
of Forestry. Durham. N.Carolina

Bulletin de l'École Nationale
Supérieure d'Agriculture de
Tunis

Bulletin de l'École Nationale
Supérieure d'Agronomie. Nancy

Bulletin of the Ecological
Society of America. Tucson

Bulletin of economics and
statistics, Government of
Gujarat. Ahmedabad

Bulletin economique, Banque
Nationale de Viet-Nam. Saigon

Bulletin. Edinburgh and East of
Scotland College of Agriculture.
Edinburgh

Bulletin of Endemic Diseases.
Baghdad

Bulletin des engrais. Paris

Bulletin of Entomological
Research. London

Bulletin of the Entomological
Society of America. Washington

Bulletin entomologique de la
Pologne. Wrocław

Bulletin of Environmental
Contamination and Toxicology.
USA

Bulletin of Environmental Control
and Toxicology. New York

Bulletin of Epizootic Diseases
of Agrica. Muguga.

Bulletin of the European
Communities. Brussels

Bulletin of the Experiment
Forests, Tokyo University of
Agriculture and Technology

Bulletin. Experiment Station of
the South African Sugar
Association. Natal

Bulletin of Experimental Animals.
Tokyo

Bulletin of the Faculty of
Agriculture, Hirosaki University. Hirosaki

Bulletin of the Faculty of
Agriculture, Kagoshima University. Kagoshima

Bulletin of the Faculty of
Agriculture, Mie University.
Tsu, Mie

Bulletin of the Faculty of
Agriculture, Shinshu University.
Ina

Bulletin of the Faculty of
Agriculture, Shizuoka University
Iwata

Bulletin of the Faculty of
Agriculture, Tamagawa University
Machida City

Bulletin of the Faculty of
Agriculture, University of
Miyazaki. Miyazaki

Bulletin of the Faculty of
Agriculture, Yamaguchi University. Shimonoseki

Bulletin of the Faculty of
Textile Fibres, Kyoto Univer-
sity of Industrial Arts and
Textile Fibers. Kyoto

Bulletin of the First Agronomy
Division, Tokai-Kinki National
Agricultural Experiment Station.
Tsu-City

Bulletin de la Fédération
Française d'Économie Montagnards

Bulletin. Fisheries Research
Board of Canada. Ottawa

Bulletin. Agricultural Experiment
Station. Inst. of Food and
Agricultural Science. University
Florida

Bulletin. Fonds de Recherches
Forestières. Universite Laval.
Quebec

Bulletin, Ford Forestry Centre.
L'Anse

Bulletin. Foreign Agricultural
Service US Department of
Agriculture, Washington

Bulletin. Forest Products Research
Laboratory. London

Bulletin. Forests Commission,
Victoria. Melbourne

Bulletin. Forests Department,
Western Australia. Perth

Bulletin of the Forestry
Commission. London

Bulletin of the Forestry
Commission, Tasmania. Hobart

Bulletin. Forestry and Timber
Bureau. Canberra

Bulletin. Geological Survey of
New Zealand. Wellington, NZ

Bulletin. Georgia Agricultural
Experiment Station. Athens

Bulletin of the Government Forest
Experiment Station, Meguro.
Tokyo

Bulletin of Grain Technology.
Hapur

Bulletin du Groupe française des
Argiles. Paris

Bulletin of the Hatano Tobacco
Experiment Station. Hatano

Bulletin. Hawaii Agricultural
Experiment Station. Honolulu

Bulletin. Highway Research Board.
(National Research Council)
Washington

Bulletin of the Hiroshima
Agricultural College. Saijo.
(Hiroshima Prefecture)

Bulletin of the History of Medicine
Baltimore

Bulletin Hoblitzelle Agricultural
Laboratory. Renner.

Bulletin of the Hokuriki
Agricultural Experiment Station.
Takada

Bulletin horticole. Liège

Bulletin. Horticultural Division,
Tokai-Kinki Agricultural
Experimental Station. Okitsu

Bulletin of the Horticultural
Research Station, (Ministry of
Agriculture and Forestry).

Series A (Hiratsuka)
  B (Okitsu)
  C (Morioka)
  D

Bulletin. I.B.E.C. Research
Institute. New York

Bulletin. Ibaraki Horticultural
Experiment Station

Bulletin. Idaho Agricultural
Experiment Station. Moscow

Bulletin. Illinois Agricultural
Experiment Station. Urbana

Bulletin of the Illinois State
Natural History Survey. Urbana

Bulletin. Indian Central Coconut
Committee. Ernakulam

Bulletin. Indian Central Jute
Committee. Calcutta

Bulletin. Indian Coffee Board
Research Department. Balehonnur

Bulletin. Indian Council of
Agricultural Research. Delhi

Bulletin of the Indian Phytopathological Society. New Delhi

Bulletin of the Indian Society for
Malaria and other Communicable
Diseases

Bulletin. Indiana Agricultural
Experiment Station. Lafayette

Bulletin. Indonesian Economic
Studies. Canberra

Bulletin d'information CORESTA.
Centre de cooperation pour les
recherches scientifiques
relatives at tabac. Paris

Bulletin d'information INEAC.
Institut national pour l'etude
agronomique du Congo belge.
Bruxelles

Bulletin d'information de la
mutualite agricoles. Paris

Bulletin d'informations
scientifiques et techniques.
Paris

Bulletin d'informations techniques
Centre technique du bois. Paris

Bulletin de l'Institut botanique.
Sofia

Bulletin de l'Institut francaise
d'Afrique Noire. Paris, Dakar

Serie A
    B

Bulletin. Institut francaise du
cafe et du cacao

Bulletin de l'Institut d'hygiène
du Maroc. Rabat

Bulletin de l'Institut de
microbiologie de l'Academie
bulgare des sciences. Sofia

Bulletin de l'Institut national
pour l'etude agronomique du
Congo belge

Bulletin de l'Institut national
de la Sante et de la recherche.
Paris

Bulletin de l'Institut oceano-
graphique. Monaco

Bulletin de l'Institut Pasteur.
Paris

Bulletin de l'Institut des
Peches maritimes du Maroc. Rabat

Bulletin. Institut de Recherches
Agronomiques de Madagascar

Bulletin de l'Institut de
recherches scientifiques du
Congo. Brazzaville

Bulletin de l'Institut r. des
sciences naturelles de Belgique.
Bruxelles

Bulletin of the Institute of
Agricultural Research, Tohoku
University. Sendai

Bulletin of the Institute of
Chemistry, Academia sinica.
Taipei

Bulletin. Institute of Development
Studies, University of Sussex.
Brighton

Bulletin of the Institute of
Marine Medicine in Gdansk.
Gdansk

Bulletin of the Institute of
Medical Research. University
of Madrid

Bulletin from the Institute of
Medical Research, Federation of
Malaya. Kuala Lumpur.

Bulletin of the Institute of
Public Health. Tokyo

Bulletin of the Institute of
Zoology, Academia Sinica. Taiwan

Bulletin of the Institution of
Mining and Metallurgy. London

Bulletin. International Association
of Scientific Hydrology. London

Bulletin, International Association
of Wood Anatomists. Zurich

Bulletin. International Institute
for Land Reclamation and
Improvement. Wageningen

Bulletin of the International
Institute of Refrigeration.
Paris

Bulletin. International North
Pacific Fisheries Commission.
Vancouver

Bulletin. Iowa Agricultural
Experiment Station. Ames

Bulletin du Jardin botanique
Nationale de Belgique.
Bruxelles

Bulletin of the Kagoshima Tobacco
Experiment Station. Kagoshima

Bulletin of the Kanagawa
Agricultural Experiment Station.

Bulletin of the Kanagawa
Horticultural Experiment
Station

Bulletin of the Kansas Agricultural
Experiment Station. Manhattan

Bulletin of the Kentucky
Agricultural Experiment Station
Lexington

Bulletin of the Kenya Sisal Board

Bulletin of the Kyoto University
Forests. Kyoto

Bulletin of the Kyushu Agricultural Experiment Station.
Hainuzukamachi., Chikugo

Bulletin. Kyushu University
Forests, Fukuoka

Bulletin. Lanbouwproefstation,
Surinam. Paramaribo

Bulletin of the Los Angeles
Neurological Society. Los
Angeles

Bulletin of the Louisiana
Agricultural Experiment Station
Baton Rouge

Bulletin de Madagascar. Tananarive

Bulletin of the Maine Agricultural
Experiment Station. Crono

Bulletin of the Maine Forest
Service. Augusta.

Bulletin of Marine Science of the
Gulf and Caribbean. Coral
Gables

Bulletin of the Maryland
Agricultural Experiment Station.
College Park

Bulletin of the Massachusetts
Agricultural Experiment Station
Amherst

Bulletin of the Mauritius
Institute. Port Lonis

Bulletin. Mauritius Sugar Industry
Research Institute. Reduit

Bulletin médical. Paris

Bulletin medical de l'A.O.F. Dakar

Bulletin et mémoires de l'École
préparatoire de médicine et de
pharmacie de Dakar. Dakar

Bulletin et mémoires de la
Société medicale des hôpitaux
de Paris

Bulletin mensuel de la Société
linnéenne de Lyon

Bulletin Mensuel. Societe de
medecine militaire française.

Bulletin mensuel de la Société
vétérinaire pratique de France.
Paris

Bulletin. Ministry of Agriculture,
Egypt. Veterinary Laboratories
and Research Administration

Bulletin. Ministry of Agriculture,
Fisheries and Food. London

Bulletin of the Ministry of
Agriculture and Lands, Jamaica

Bulletin. Ministry of Agriculture,
Lands and Fisheries, Barbados

Bulletin. Ministry of Food and
Agriculture, Ghana

Bulletin. Minnesota Agricultural
Experiment Station. St. Paul

Bulletin of the Mississippi
Agricultural Experiment Station.
State College

Bulletin. Missouri Agricultural
Experiment Station. Columbia

Bulletin of the Montana
Agricultural Experiment Station.
Bozeman

Bulletin. Montana Forest and
Conservation Experiment Station,
Montana State University, Missoula

Bulletin of the Morioka Tobacco
Experiment Station

Bulletin de la Murithienne.
Société valaisanne des sciences
naturelles. Sion.

Bulletin of the Museum of
Comparative Zoology at Harvard
College. Cambridge, Mass.

Bulletin du Museum national
d'histoire naturelle. Paris

Bulletin. Mushroom Growers
Association. UK

Bulletin of the Naniwa University.
Sakai.

Series B. Agricultural and
natural sciences

Bulletin of the National Institute
of Agricultural Sciences. Tokyo

Ser. A. Physics and Statistics
B. Soils and Fertilizers
C. Plant Pathology, (after-
wards Phytopathology)
and Entomology
D. Plant Physiology,
Genetics and Crops
G. Animal Husbandry
H. Farm Management and Land
Utilization

Bulletin of the National Institute
of Animal Industry. Tokyo

Summaries of Reports

Bulletin of the National Institute
of Hygienic Sciences. Tokyo

Bulletin of the National Institute
of Sciences of India. New Delhi

Bulletin of the National Museum,
State of Singapore. Singapore

Bulletin of the National Research
Council. Washington

Bulletin of the National Science
Museum. Tokyo

Bulletin of the National Society
of India for Malaria and other
Mosquito Bourne Disease. Delhi

Bulletin. National and University
Institute of Agriculture,
Rehovot

Bulletin of the Nebraska Agricultural Experiment Station. Lincoln

Bulletin of the Nevada Agricultural Experiment Station. Reno

Bulletin of the New Hampshire
Agricultural Experiment Station.
Durham

Bulletin of the New Jersey Academy
of Science

Bulletin of the New Jersey
Agricultural Experiment Station.
New Brunswick

Bulletin of the New Mexico
Agricultural Experiment Station.
State College, New Mexico

Bulletin of the New York Academy
of Medicine. New York

Bulletin of the New York State
Agricultural Experiment Station.
Geneva

Bulletin of the New York College
of Forestry. Syracuse

Bulletin of the New York State
Flower Growers Incorporated.
New York

Bulletin of the New York State
Museum and Science Service.
Albany

Bulletin of the New Zealand
Department of Scientific and
Industrial Research. Wellington

Bulletin. New Zealand Meat and
Wool Board's Economic Service.
Wellington

Bulletin of the Nigerian Forestry
Departments. Ibadan

Bulletin. Nihon University College
of Agriculture and Veterinary
Medicine. Japan

Bulletin of the Nippon Agricultural Research Institute. Tokyo

Bulletin of the Nippon Veterinary and Zootechnical College. Japan

Bulletin of the North Carolina Agricultural Experiment Station. West Raleigh

Bulletin of the North Carolina Flower Growers.

Bulletin of the North Dakota Agricultural Experimental Station. Fargo

Bulletin. North of Scotland College of Agriculture. Aberdeen.

Bulletin. Office international des epizooties. Paris

Bulletin de l'Office international de la vigne. Paris

Bulletin de l'Office international du vin. Paris

Bulletin of the Ohio Florists Association

Bulletin. Oji Institute for Forest Tree Improvement. Kuriyama, Hokkaido

Bulletin of the Oklahoma Agricultural Experiment Station. Stillwater

Bulletin of the Ontario Department of Agriculture

Bulletin of the Oregon Agricultural Experiment Station. Corvallis

Bulletin. Oregon State University Forest Research Laboratory. Corvallis

Bulletin of the Osaka Medical School. Osaka

Bulletin. Otago Catchment Board

Bulletin of the Oxford University Institute of Economics and Statistics. Oxford

Bulletin of the Pennsylvania Agricultural Experiment Station. State College, Pa.

Bulletin of the Pennsylvania Flower Growers.

Bulletin of the Pharmaceutical Research Institute, Takatsuki. Osaka

Bulletin of the Porto Rico Agricultural Experiment Station. Washington

Bulletin des Recherches Agronomiques de Gembloux

Bulletin of Regional Research Laboratory, Jammu. Council of Scientific and Industrial Research

Bulletin. Research Department, Indian Coffee Board. Ballehonur

Bulletin. Research and Information Division, Farm Credit Administration. Washington

Bulletin. Research Institute for agricultural economics. Budapest

Bulletin of the Research Institute for Food Science. Kinki University. Osaka

Bulletin of the Research Institute for Food Science, Kyoto University. Kyoto

Bulletin of the Research Institute of the Sumatra Planters' Association. Medan

Bulletin of the Reserve Bank India, Bombay

Bulletin. Rhode Island Agricultural Experiment Station. Kingston

Bulletin of Rural Economics and Sociology. Ibadan

Bulletin, School of Forestry, Montana State University. Missoula

Bulletin, School of Forestry, Stephen F. Austin State College, Nacogdoches. Texas

Bulletin der Schweizerischen Akademie der medizinischen Wissenschaften. Basel

Bulletin scientifique. Conseil des academies de la RPF Yugoslavie Beograd, etc.

Bulletin scientifique. Ministere de la France d'outre mer

Bulletin of the Schottish Milk
Marketing Board, Glasgow

Bulletin des séances. Académie r.
des sciences coloniales (d'outre
mer). Bruxelles.

Bulletin des séances. Institut r.
colonial belge. Bruxelles

Bulletin of the Sericultural
Experiment Station, Japan.
Tokyo

Bulletin du Service de la Carte
Geologique d'Alsace et de
Lorraine. Strasbourg

Bulletin du Service de Culture et
d'Etudes du Peñplier et du
Saule, Paris

Bulletin of the Shikoku Agricultural Experiment Station.
Zentsuji

Bulletin of the Shimane Agricultural College. Matsue

Bulletin of the Shizuoka Prefectural Citrus Experiment Station.

Bulletin signaletique. Centre
national de la recherche
scientifique. Paris

Bulletin de la Société belge
d'ophtalmologie. Bruxelles

Bulletin. Société botanique de
France. Paris

Bulletin de la Société de chimie
biologique. Paris

Bulletin des sociétés chimiques
belges. Brussels

Bulletin de la Société chimique
de France. Paris

Bulletin. Société entomologique
d'Egypte. La Caire

Bulletin de la Société entomologique de France

Bulletin de la Société entomologique de Mulhouse

Bulletin de la Société française
de dermatologie et de syphiligraphie. Paris

Bulletin de la Société française
de mineralogie (et de
cristallographie). Paris

Bulletin de la Société française
de physiologie végétale. Paris

Bulletin de la Société géologique
de France. Paris

Bulletin de la Société d'histoire
naturelle de l'Afrique du Nord.
Alger

Bulletin de la Société d'histoire
naturelle du Doubs. Besançon

Bulletin de la Société d'histoire
naturelle de Toulouse

Bulletin de la Société medicale
d'Afrique noire de langue
française. Dakar

Bulletin de la Société nationale
d'Horticulture de France

Bulletin de la Société neuchâteloise des sciences naturelles.
Neuchâtel

Bulletin. Société d'ophtalmologie
d'Egypte. Alexandrie and
Le Caire

Bulletin de la Société d'ophtalmologie de France. Paris

Bulletin de la Société portugaise
des sciences naturelles. Lisbon

Bulletin de la Société de·
pathologie exotique et de ses
Filiales. Paris

Bulletin. Société r. de botanique
de Belgique. Bruxelles

Bulletin. Société r. forestiere
de Belgique. Bruxelles

Bulletin de la Société r. de
zoologie d'Anvers. Anvers

Bulletin de la Société des
sciences et des lettres de Łódź.
Łódź.

   Classe III. Sciences mathematiques et naturelles

Bulletin de la Société des
sciences naturelles et
physiques du Maroc. Rabat.

Bulletin de la Société des
sciences vétérinaires de Lyon.
Lyons

Bulletin de la Société des
sciences vétérinaires et de
médecine comparée de Lyon

Bulletin de la Société scientifique
d'hygiène alimentaire. Paris

Bulletin de la Société vaudoise
des sciences naturelles. Lausanne

Bulletin de la Société vétérinaire
hellenique. Athene

Bulletin de la Société vétérinaire
de zootechnie d'Algerie. Alger

Bulletin de la Société zoologique
de France. Paris

Bulletin of the Society of
Entomology, Taiwan Provincial
Chung-Hsing University

Bulletin. Soil Bureau, DSIR.
New Zealand. Wellington

Bulletin. Soil Survey of Great
Britain. Harpenden

Bulletin. South Africa (Republic
of) Department of Agricultural
Technical Services. Pretoria

Bulletin of the South Carolina
Agricultural Experiment Station.
Clemson College

Bulletin. South Dakota Agricultural
Experiment Station. Brookings

Bulletin of the Southern California
Academy of Sciences. Los Angeles

Bulletin. State of Connecticut.
State Geological and Natural
History Survey

Bulletin of the State Institute
of Marine and Tropical Medicine
in Gdansk

Bulletin. State of Israel
Agricultural Research Station.
Beit-Dagan

Bulletin. State of Israel
Ministry of Agriculture. Soil
Conservation Division

Bulletin. State Plant Board of
Florida. Gainesville

Bulletin of the Stoneham Museum,
Kitale, Kenya Colony. Nairobi

Bulletin. Storrs Agricultural
Experiment Station, Storrs. Conn

Bulletin of Sugar Beet Research
Japan Sugar Beet Improvement
Foundation. Tokyo

Bulletin. Swaziland Department of
Agriculture. Mbabane

Bulletin. Taiwan Agricultural
Research Institute. Taipei

Bulletin. Taiwan Forestry
Research Institute. Taipei

Bulletin. Tanganyika Ministry of
Agriculture. Dar-es-Salaam

Bulletin. Tea Research Institute
of Ceylon. Talawakella

Bulletin Technique. Divisions des
Sols. Province de Quebec.
Ministere de l'Agriculture.
Quebec

Bulletin technique d'information
des ingénieurs des services
agricoles. Paris

Bulletin Technique d'information
Ministere de l'Agriculture,
Paris

Bulletin of the Tennessee
Agricultural Experiment Station
Knoxville

Bulletin. Texas Agricultural
Experiment Station.
College Station

Bulletin. Texas Agricultural and
Mechanical College Extension
Service. College Station

Bulletin of the Texas Research
Foundation

Bulletin. Tobacco Research Board
of Rhodesia and Nyasaland.
Salisbury

Bulletin of the Tohoku National
Agricultural Experimental
Station. Morioka

Bulletin of the Tokai-Kinki
National Agricultural Experiment
Station. Tsu, Kanaya

Bulletin of Tokyo Medical and
Dental University. Tokyo

Bulletin of the Tokyo University
Forests. Tokyo

Bulletin of the Torrey Botanical
Club. New York

Bulletin of the Tottori University Forests. Tottori

Bulletin trimestriel de la Societe mycologique de France. Paris

Bulletin of the Tulane Medical Faculty. New Orleans

Bulletin. United Planters' Association of Southern India. Coimbatore

Bulletin. United States National Museum. Smithsonian Institution. Washington

Bulletin. University of Alberta. Edmonton

Bulletin. University of Massachusetts. College of Agriculture. Amherst

Bulletin of University of Osaka Prefecture. Sakai, Osaka.

Ser.B. Agriculture and biology

Bulletin. University of Puerto Rico Agricultural Experiment Station. Rio Piedras

Bulletin of the Utah Agricultural Experiment Station. Logan

Bulletin. Utah State University, Logan, Utah

Bulletin. Utsŏnomiya University. (Forests)

Bulletin of the Vermont Agricultural Experiment Station. Burlington

Bulletin of the Veterinary Institute in Pulawy. Poland

Bulletin of the Virginia Agricultural Experiment Station. Blacksburg

Bulletin. Virginia Polytechnic Institute Engineering Experiment Station. Blacksburg

Bulletin. Výzkumny Ústav Zelinářský. Olomouc.

Bulletin of the Washington Agricultural Experiment Station. Pullman

Bulletin of the West Virginia University Agricultural Experiment Station. Morgantown.

Bulletin of the Wildlife Disease Association. USA

Bulletin. Wisconsin Agricultural Experiment Station. Madison

Bulletin. Woods and Forests Department. South Australia. Adelaide

Bulletin. Wood Research Laboratory, Virginia Polytechnic Institute. Blacksburg

Bulletin of the World Health Organization. Geneva

Bulletin. Wyoming Agricultural Experiment Station. Laramie

Bulletin. Wyoming Game and Fish Commission. Laramie

Bulletin. Yale University, School of Forestry. New Haven

Bulletin of the Yamagata Agricultural College. (Agricultural Science) Tsuruoka

Bulletin. Yamanashi Prefectural Forest Experiment Station. Fuji-Yoshida

Bulletin of Zoological Nomenclature London

Bulletin of the Zoological Society, College of Science, Nagpur. India

Bulletin. Zoological Society of Egypt.

Bumazhnaya promyshlennost'. Moskva

Búnaŏarrit. Reykjavik

Bundesversuchsinstitut für Kulturtechnik und Technische Bodenkunde. Petzenkirchen

Burma Medical Journal. Rangoon

Business and Government Review. Colombia. Tennessee

Buskap og avdrått. Tidsskrift for svin- og storfehold. Gjøvik

Butler University Botanical Studies. Indianapolis

Butter Fat and Solids. Sydney

Butter-Fat. Vancouver

Byulleten' éksperimental'noĭ
biologii i meditsiny. Moskva

Byulleten' Glavnogo botanicheskogo
sada. Leningrad

Byulleten' Instituta biologii
vodokhranilishch. Moskva.

Byulleten' Kirgizskogo nauchno-
issledovatel'skogo Institute
Zamledeliya. Frunze

Byulleten' Moskovskogo obshchestva
ispytatelei prirody. Moskva.

Otdel Biologicheskii

Byulleten' nauchno-tekhnicheskoĭ
informatsii po sel'skokhozyaĭst-
vennoĭ mikrobiologii. Leningrad

Byulleten' nauchno-tekhnicheskoĭ
informatsii vsesoyuznogo
instituto gel'mintologii im.
K.I. Skryabina.

Byulleten' sovetskoi antarktich-
eskoĭ ekspeditsii

C

C.C.H. Mykologicky sbornik
v Praze

C.M.I. Descriptions of pathogenic
fungi and bacteria. Kew

C.S.I.R. Research Review. South
African Council of Scientific
and Industrial Research.
Pretoria

C.S.I.R.O. Abstracts. Melbourne

C.S.I.R.O. Plant Introduction
Review. Australia

C.S.I.R.O. Wildlife Research.
Melbourne

Cacao atualidades.

Cacao. Inter-American Cacao Centre
Turrialba

Cactus and Succulent Journal of
Great Britain. Elstree

Café. Turrialba

Café-cacao-thé. Nogent-sur-Marne

Cahiers d'agriculture des pays
chauds. Paris

Cahiers de biologie marine.
Roscoff

Cahiers du Centre d'etudes et de
recherches économiques et
sociales. Université Tunis
(Série géographique). Tunis

Cahiers du Centre National des
Exposition et concours agricoles
Paris

Cahiers. Centre technique du bois.
Paris

Cahiers économiques de Bruxelles.
Bruxelles

Cahiers économiques et sociaux.
Kinshasa

Cahiers d'Etudes Africaines.
Rijswijk

Cahiers des ingénieurs agronomes.
Paris

Cahiers de Institut Economique
Agricole, Ministere de
l'Agriculture. Bruxelles

Cahiers de Institut National de
Gestion et d'Economie Rurale
(Supplement a la revue Organis-
ation et Gestion de l'Entreprise
Agricole). Paris

Cahiers de l'Institut de Science
economique appliquee. Serie AG
"Progres et Agriculture". Paris

Cahiers de l'Institut des Sciences
Economiques et Sociales de
l'Universite de Fribourg Suisse.
Fribourg

Cahiers Internationaux de
Sociologie. Paris

Cahiers de la Maboké. Organe de
la Station Experimentale du
Museum National d'Histoire
Naturelle en Republique
Centrafricaine. Paris

Cahiers médicaux de l'Union
française. Alger

Cahiers de médecine vétérinaire.
Paris

Cahiers des naturalistes. Paris

Cahiers de Nutrition et de
Diététique. Paris

Cahiers d'Outre-Mer. Bordeaux

Cahiers de pédologie. Office de la Recherche scientifique et technique Outre-Mer. Paris

Cahiers de physiologie des plantes tropicales cultivees. Office de la Recherche scientifique et technique Outre-Mer. Paris

Cahiers de la recherche agronomique. Rabat

Cahiers ruraux. Bruxelles

Calcified Tissue Research. Germany

Calcutta Statistical Association Bulletin. Calcutta

Caldas medico. Manizales. Colombia

Calendar of the School of Agriculture. Aberdeen

California Agriculture. Berkeley

California Avocado Society Yearbook

California Citrograph. Los Angeles

California Fish and Game. Sacramento

California Forestry and Forest Products. California Forest Products Lab. Berkeley

California Medicine. San Francisco

California Vector Views. Berkeley

California Veterinarian. San Francisco

Camellia Journal

Campo. Seville

Campo. Tucuba

Canada Agriculture. Ottawa

Canadian Agricultural Engineering. Guelph

Canadian Bulletin on Nutrition. Ottawa

Canadian Entomologist. Ottawa

Canadian Farm Economics, Ottawa

Canadian Field Naturalist. Ottawa

Canadian Food Industries. Gardenvale. Quebec

Canadian Forest Industries. Don Mills

Canadian Geographer. Manotick

Canadian Geographical Journal. Montreal

Canadian Insect Pest Review. Ottawa

Canadian Journal of Agricultural Economics. Toronto

Canadian Journal of Agricultural Science. Ottawa

Canadian Journal of Animal Science. Ottawa

Canadian Journal of Biochemistry. Ottawa

Canadian Journal of Botany. Ottawa

Canadian Journal of Chemistry. Ottawa

Canadian Journal of Comparative Medicine. Quebec

Canadian Journal of Earth Sciences. National Research Council (Canada). Ottawa

Canadian Journal of Genetics and Cytology. ·Ottawa

Canadian Journal of Medical Sciences. Ottawa

Canadian Journal of Medical Technology. Hamilton

Canadian Journal of Microbiology. Ottawa

Canadian Journal of Plant Science. Ottawa

Canadian Journal of Physics. Ottawa

Canadian Journal of Physiology and Pharmacology. Ottawa

Canadian Journal of Public Health. Toronto

Canadian Journal of Research. Ottawa.

Section D. Zoological Sciences

Canadian Journal of Soil Science. Ottawa

Canadian Journal of Technology. Ottawa

Canadian Journal of Zoology. Ottawa

Canadian Medical Association
Journal. Toronto

Canadian Plant Disease Survey.
Ottawa

Canadian Services Medical Journal.
Ottawa

Canadian Spectroscopy.
Lachine. Quebec.

Canadian Veterinary Journal.
Ottawa

Cancer. Philadelphia

Cancer Research. Baltimore, etc.

Candollea. Organe du Conservatoire
et du Jardin botaniques de la
ville de Genève.

Cane Growers' Quarterly Bulletin.
Brisbane

Canterbury Chamber of Commerce
Agricultural Bulletin

Carbohydrate Research. Amsterdam

Cardiologia. Basel

Caribbean Agriculture. Puerto Rico

Caribbean Medical Journal.
Port of Spain

Caries Research. Basle

Carinthia II. Mitteilungen des
Naturhistorischen Landesmuseums
für Kärnten. Klagenfurt

Cartel. London

Caryologia. Giornale de citologia,
citosistematica e citogenetica.
Torino. etc.

Časopis československých veterinářů
v Brne

Časopis lékařů českých. v Praze

Časopis pro mineralogii a
geologii. Praha

Časopis Moravského musea v Brne.
Brno

Časopis Narodniho musea. Praha

Časopis Slezskeho Muzea, Opava
(Ser. C, Dendrologie)

Castanea. The Journal of the
Southern Appalachian Botanical
Club. Morgantown

Cat Fancy

Cawthron Institute. Memorial
Lecture. Nelson. N.Z.

Cawthron Institute Monographs.
Nelson. N.A.

Ceiba. A scientific journal issued
by the Escuela agricola pan-
america. Tegucigalpa

Cellule. Lierre, etc.

Cellulosa e carta. Roma

Celuloza şi Hirtie. Bucureşti

Cenicafé. Chinchina.

Cento Conference on Land Class-
ification for Non-irrigated Lands
(1966) Ankara

Central African Journal of Medicine
Salisbury

Central Asian Review. London

Centre Technique Interprofessionel
des Oléagineux Métropolitains.
Paris

Cereal Chemistry. St. Paul, Minn.

Cereal News. Ottawa

Cereal Science Today. Minneapolis

Ceres. London

Ceres. Revista bi-mensal de
divulgação de ensinamentos
teóricos e prácticos sobre
agricultura, veterinária,
industrias rurais. Minas Gerais.

Česká mykologie. Praha

Československá biologie. Praha

Československá dermatologie. Praha

Československá epidemiologie,
mikrobiologie, imunologie.
Praha

Československá farmacie. Praha

Československá hygiena. Praha

Československá parasitologie. Praha

Československá pediatrie. Praha

Československý kras

Ceylon Coconut Quarterly. Lunnwila

Ceylon Coconut Planters' Review.
Lunuwila

Ceylon Forester. Colombo

Ceylon Journal of Medical Science. Colombo

Ceylon Journal of Science. Colombo. Section B. Zoology

Ceylon Journal of Science. Colombo. Biological Sciences

Ceylon Medical Journal. Colombo

Ceylon Trade Journal. Colombo

Ceylon Veterinary Journal. Colombo, Peradeniya

Chacaras e quintaes. São Paulo

Chacra. Revista mensual de agricultura, ganaderia e industria. Buenos Aires

Chambres d'agriculture

Champignon

Champignoncultuur

Chartered Surveyor. London

Chemia analityczna. Warszawa

Chemical Abstracts. Easton, Pa.

Chemical and Engineering News. Easton, Pa. etc.

Chemical Reviews. Baltimore, etc.

Chemický průmysl. Praha

Chemie analytique. Paris

Chemie der Erde. Beitrage zur chemischen Mineralogie, Petrographie u. Geologie. Jena

Chemisch weekblad. Amsterdam

Chemische Berichte. Heidelberg und Berlin, etc.

Chemische Rundschau für Mitteleuropa und den Balkan. Budapest

Chemist-Analyst. J.T. Barker Chemical Co. Philipsburg, NJ.

Chemistry in Britain. London

Chemistry and Industry. London

Chemistry and Life. Tokyo

Chemotherapia. Basel, New York

Chesapeake science. Solomons

Chest Disease Index and Abstracts including Tubercolosis. London

Chiengmai Medical Bulletin

Child Development. Baltimore, etc.

Chimia. Zurich

Chimica. Milano

Chimica el'industria. Milano

Chimie et industrie. Paris

China Horticulture

China Quarterly, London

China Report, New Delhi

Chinese-American Joint Committee on Rural Reconstruction. Taipei

Plant industry series

Chinese Economic Studies, White Plains, NY

Chinese Journal of Ophthalmology

Chinese Journal of Surgery

Chinese Medical Journal. Peking

Chinese Medical Journal. Free China Edition. Taiwan

Chinese Veterinary Journal

Chirigaku Kenkyu Hokoku. Tokyo

Chirurg. Zeitschrift fur alle Gebiete der operativen Medizin. Berlin

Chirurgia italiana. Belluno, etc.

Chirurgia degli organi di movimento. Bologna

Chmelarstvi. Praha

Chromosoma. Zeitschrift für Zellkernund Chromosomenforschung. Berlin, Wien.

Chromosome Information Service. Tokyo

Chromatographic Reviews. Amsterdam

Chronica botanica. Leiden, Waltham, Mass.

Chronica Horticulturae

Chronique d'Actualité. Paris

Chrysanthème. Lyon

Chung Chi Journal. Hong Kong

Ciencia. Revista hispano-americana de ciencias puras y aplicados. Mexico

Ciencia e investigación. Buenos Aires

Ciencia y naturaleza. Quito

Ciencia veterinaria. Madrid

Ciencias. Anales de la Asociación española para el progreso de las ciencias. Madrid.

Ciencias politicas y sociales. Mexico

Circolare. Occervatorio per le malattie delle piante. Catanzaro

Circular. Agricultural Experiment Station, University of Georgia. Athens.

Circular. Alabama Agricultural Experiment Station. Auburn

Circular. Auburn University Agricultural Experiment Station

Circular. California Agricultural Experiment Station. Berkeley

Circular. California Agricultural Extension Service, California University. Berkeley

Circular. Clemson Agricultural College Extension Service. Clemson

Circular. College of Agriculture Cooperative Extension Service, University of Illinois. Urbana

Circular. Connecticut Agricultural Experiment Station. New Haven

Circular. Delaware Agricultural Experiment Station

Circular. Estación experimental agricola, Tucumán. Buenos Aires

Circular. Fish and Wildlife Service. United States Department of Agriculture. Washington

Circular. Florida University Agricultural Experiment Station. Gainesville

Circular. Georgia Agricultural Experiment Station. Athens

Circular. Georgia College of Agriculture

Circular. Georgia Coastal Plain Experiment Station. Tifton

Circular. Hawaii Agricultural Experiment Station. Honolulu

Circular. Indiana Agricultural Experiment Station. Lafayette

Circular. Illinois Agricultural Experiment Station. Urbana

Circular. Illinois Natural History Survey. Urbana

Circular. Instituto agronómico do Norte. Belem

Circular do Instituto agronomico do Sul

Circular. Instituto de pesquisas e experimentação agropecuarias do Norte. Belem

Circular do Instituto de pesquisas e experimentação agropecuarias do Sul

Circular. Kansas Agricultural Experiment Station. Manhattan

Circular. Kentucky Agricultural Experiment Station. Lexington

Circular. Louisiana Agricultural Experiment Station. Baton Rouge

Circular. Mississippi Agricultural Experiment Station. State College

Circular. Montana Agricultural Experiment Station. Bozeman

Circular. New Jersey Agricultural Experiment Station. New Brunswick

Circular. North Dakota Agricultural College Extension Service

Circular. Oklahoma Agricultural Experiment Station. Stillwater

Circular. Pennsylvania Agricultural Experiment Station. State College

Circular. Porto Rico Agricultural Experiment Station

Circular. South Carolina
Agricultural Experiment Station
Clemson College

Circular. South Dakota Agricultural
Experiment Station. Brookings

Circular. Tennessee Agricultural
Experiment Station. Knoxville

Circular. Texas A & M College
Extension Service

Circular. US Department of
Agriculture. Washington

Circular. University of Illinois.
College of Agriculture.
Cooperative Extension Service
Urbana

Circular. Washington Agricultural
Experiment Station. Pullman.

Circular of the Wisconsin University of Agriculture Extension
Service. Madison

Circular. Wyoming Agricultural
Experiment Station. Laramie

Circulation. Journal of the
American Heart Association.
New York

Circulation Research. New York

Cirugia. Madrid

Cirugia y ciruganos. Mexico

Citologija

Citrus Industry. Tampa, etc.

Citrus Leaves. Los Angeles

Citrus and Vegetable Magazine

Civil Engineering and Public
Works Review. London

Civilisations. Bruxelles

Clay Minerals. Galashiels, London

Clay Minerals Bulletin.
Galashiels, London

Clay Science. Tokyo

Clays and Clay Minerals.
Proceedings of the National
Conference on Clays and Clay
Minerals. Washington, London

Clemson College of Agricultural
Research

Cleveland Clinic Quarterly.
Cleveland

Clinica chimica acta. Amsterdam,
London

Clinica y laboratorio. Zaragoza

Clinica nuova. Roma

Clinica pediatrica. Modena,
Bologna

Clinica terapeutica. Roma

Clinica Veterinaria. Italy

Clinica Veterinaria e rassegna di
polizia sanitaria e di igiena.
Milano

Clinical Chemistry. Baltimore,
New York

Clinical and Experimental
Immunology. UK

Clinical Neurology Japan

Clinical Pediatrics. New York

Clinical Pharmacology and
Therapeutics. St. Louis

Clinical Science, incorporating
Heart. London

Cocoa Growers' Bulletin.
Bournville

Coconut Bulletin issued by the
Indian Central Coconut Committee
Ernakulam

Coconut Sitatuion. FAO Rome

Coedwigwr. Magazine of the
Forestry Society of North Wales

Coffee. Turrialba

Cold Spring Harbor Symposia on
Quantitative Biology. Cold
Spring Harbor

Collana verde Ministero dell'
Agricoltura e delle Foreste,
Direzione generale dell'
Economia e delle Foreste, Roma

Collectanea botanica a Barcinonensi
Botanico Instituto edita.
Barcinone

Collected Papers. Institute of
Animal Physiology, Barbraham

Collected Papers. Macaulay
Institute for Soil Research.
Craigiebuckler

Collected Papers. Rowett Research
Institute

Collected Papers. School of
Veterinary Medicine, University
of Cambridge

Collection of Czechoslovak
Chemical Communications.
English Edition. Praha

Collection Droit de la Coopération
Économiques et Sociale
Internationale. Paris

Collection Marchés et Structures
Agricoles, Paris

Colloque International sur la
Retention et la migration des
ions radioactifs dans les sols.
Gif-sur-Yvette

Colloque Internationale Termites
Africa. UNESCO

Colloques Internationaux du
Centre National de la Recherche
Scientifique. Paris

Colombia Today. New York

Colonial Research Studies.
Department of Technical
Cooperation. London

Colorado Farm and Home Research.
Fort Collins

Commercial Fisheries Review.
Fish and Wild Life Service.
Washington

Commercial Grower. London

Commission International Federation
of Surveyors. London

Committee Paper of the Sugar Beet
Research and Education Committee
Plant Breeding Institute.
Cambridge

Commodity Bulletin Series,
FAO. Rome

Commodity Reference Series.
FAO. Rome

Commodity Report. Bureau of
Agricultural Economics. Canberra

Common Market Law Review. Leyden

Commonwealth Agriculturist.
Melbourne

Commonwealth Forestry Review.
London

Commonwealth Phytopathological
News. London

Commonwealth Survey. UK

Communicación, Instituto Forestal
de Investigaciones y Experiencias
Madrid

Communication. Department of
Agricultural Research of the
Royal Tropical Institute.
Amsterdam

Communication. Institute of
Forestry Research, Agricultural
University. Wageningen

Communicationes. Institutus
Forestalis Fenniae. Helsinki

Communicationes Instituti
forestalis cechosloveniae
(Vyzkumny Ustav Lesniho
Hospodarstvi a Myslivosti)
Zbraslav-Stradney.

Communicationes Veterinariae.
Bogor

Community Development Journal.
Manchester

Comparative Biochemistry and
Physiology. London

Comparative studies in society and
history. Den Haag

Compost Science. Emmaus

Comptes rendus de l'Académie
d'Agriculture de France. Paris

Comptes rendue de l'Académie
bulgare de sciences. Sofia

Comptes rendue de l'Académie des
sciences de Paris

Comptes rendus du 4e Congres
International des Algues Marines
Biarritz. Pergamon Press

Compte rendu d'activité, Association Foret-Cellulose (AFOCEL
Paris

Compte rendu du Congrès des
sociétés savantes de Paris
et des départments. Section des
sciences. Paris

Compte rendu hebdomadaire des
séances de l'Académie d'agriculture de France. Paris

Compte rendu hebdomadaire des
séances de l'Académie des
sciences. Paris Series A,B,C,D

Compte rendu des journées d'études
sur les Herbicides (Columa).
Paris

Compte rendu de 3e Potassium
colloquium de l'Institut
International de Potasse, Lisbon

Compte rendu de recherches.
Institut pour l'encouragement
de la recherche scientifique
dans l'industrie et l'agriculture
Bruxelles

Compte rendu des seances
mensuelles. Societe des
sciences naturelles et
physiques du Maroc. Rabat

    Section de Pedologie

Comptes rendus des séances de la
Société de biologie et de ses
Filiales. Paris

Compte rendu de la Société des
sciences et des lettres de
Wrocław. Wrocław

Compte rendu sommaire des
séances. Société de
biogeographie. Paris

Compte rendu sommaire des
seances de la Societe
geologique de France. Paris

Compte rendu des travaux du
Laboratoire de Carlsberg.
Copenhague

Computer Bulletin. London

Computers and Biomedical Research
USA

Comunicação. Missao de Estudos
Agronómicos do Ultramar. Lisboa

Comunicaciones del Instituto
nacional de investigación de las
ciencias naturales y museo
argentino de ciencias naturales
"Bernardino Rivadavia". Buenos
Aires.

    Ciencias Zoológicas
Comunicaciones zoológicas del
Museo de historia natural de
Montevideo

Comunicările Academiei republicii
populare romîne. Bucureşti

Comunicările de botanică.
Bucureşti

Concours médical. Paris

Condor. Santa Clara, Cal.

Conferencias. Instituto nacional
de investigaciones agronómicas.
Madrid

Congo-Afrique. Kinshasa

Congrès pomologique. Société
pomologique de France.
Villefranche-sur-Saône

Coniglicoltura. Iraly

Conjuntura economica.
Rio de Janeiro

Connecticut State Medical Journal.
New Haven

Conselhos para a defsea sanitaria
das culturas. Direccao-Geral
dos Servicos agricolas. Portugal

Conserva. Den Haag

Conservation Research Report.
US Department of Agriculture,
Agricultural Research Service.
Washington

La Conserve agricole, Paris

Consommation. Paris

Contemporary Japan. Tokyo

Contemporary Review. London

Contributi. Istituto di ricerche
agrarie. Milano

Contributi scientifico-pratici per
una miglione conoscenza ed
utilizzazione del legno.
Istituto Nazionale del Legno.
Florence

Contributii botanice.
Universitatea "Babes-Bolyai"
din Cluj

Contributions. Boyce Thompson
Institute (for Plant Research)
Menasha. Wis.

Contributions from the Brooklyn
Botanic Garden. Brooklyn

Contribution. Canada Department of
Forestry. Forest Research Branch
Ottawa

Contribution. The Committee on
Desert and Arid Zones Research
Southwestern and Rocky Mountain
Division, A.A.A.S. New Mexico
State University. Univ. Park

Contributions on Eucalypts in
Israel II. National and Univer-
sity Institute of Agriculture.
Ilanot

Contributions. Fonds de recherches
forestieres de l'Université
Laval. Québec

Contributions de l'Institut
botanique de l'Université de
Montréal

Contribution, Institute of Forest
Products, University of
Washington College of Forest
Resources, Seattle

Contributions from the New South
Wales National Herbarium.
Sydney

Contributions from the United
States National Herbarium.
Washington

Converting Industry. London

Coolia. Contactblad van de
Nederlandse mycologische
vereniging. Leiden

Cooperation et Developpement.
Paris

Co-operative Economic Insect
Report. Bureau of Entomology and
Plant Quarantine, US Department
of Agriculture, Washington

Cooperative Extension Publication,
Oregon State University,
Corvallis. Ore

Cooperative Extension Service
Circular. New Mexico State
University. Las Cruces. N.Mexico

Copeia. New York, etc.

Cornell Extension Bulletin.
Ithaca

Cornell International Agricultural
Development Bulletin

Cornell International Agricultural
Mimeograph. Department of
Agricultural Economics. New York
State College of Agriculture,
Cornell University. Ithaca.

Cornell Plantations. Ithaca

Cornell Veterinarian. Ithaca

Corrosion. National Association
of Corrosion Engineers.
Houston. Tex.

Cost Accountant. London

Coton et fibres tropicales.Paris

Cotton Growing Review. London

Country Landowner. London

Courrier. Paris

Courrier de l'Institut d'Hygiene
Belgrade

Cronache economiche. Torino

Crop Science. Madison, Wisconsin

Crops and Soils. Madison

Crybiology. Rockville, Md.

Cuadernos Americanos, Mexico, DF

Cuadernos del Instituto de
Investigaciones Cientificas,
Universidad de Nuevo Leon, Mexico

Cukoripar. Budapest

Cultured Dairy Products Journal.
Ithaca. NY

Cultuur en Handel. Kortenberg

Bultuurtechniek. Rotterdam

Cuprum pro Vita. Transactions of a
Symposium. Vienna

Current Background. Hong Kong

Current Contents. Philadelphia

Current Food Additives Legislation,
FAO Rome

Current Report. West Virginia
Agricultural Experiment Station,
Morgantown, W.Va.

Current Science. Bangalore

Current Therapeutic Research,
Clinical and Experimental.
New York

Current Topics in Microbiology
and Immunology. Germany

Curtis's Botanical Magazine.
London

Cyanamid Crop Protection Bulletin

Cyprus Medical Journal. Nicosia

Cytogenetics. Switzerland/USA

Cytologia. Tokyo

Czechoslovak Economic Papers.
Prague

D

D.C.K. Information

Daffodil and Tulip Year Book.
London

Dairy Farmer. Ipswich.

Dairy Farming Annual.
Palmerston North

Dairy Herd Improvement Letter. USA

Dairy and Ice Cream Field.
New York
(Formerly Ice Cream Field and
Ice Cream Trade Journal)

Dairy Industry Journal of
Southern Africa. Pretoria

Dairy Industries. London

Dairy Research Review Series
An Foras Taluntais. Dublin

Dairy Science Abstracts.
Shinfield

Dairy Shorthorn Journal. London

Dairy Situation. Canberra

Dairyfarming Annual. Massey
Agricultural College. Palmerston
North, N.Z.

Dairyfarming Digest. Melbourne

Dalgety-N.Z.L. Annual Wool Digest
Sydney

Danish Foreign Office Journal.
Copenhagen

Danish Medical Bulletin.
Copenhagen

Danish Review of Game Biology.
Copenhagen

Dansk botanisk Arkiv. Kjøbenhavn

Dansk Dendrologisk Årsskrift.
Copenhagen

Dansk Landbrug. Køkbenhavn

Dansk Maanedskrift for Dyrlaeger.
Køkbenhavn

Dansk pelsdyravl. Køkbenhavn

Dansk Skovforeningens Tidsskrift.
Køkbenhavn

Dári de Seamá ale Şedintelor,
Republica Populara Románá.
Comitetul Geologic. Bucureşti.

Darviniana. Revista del Instituto
de botanica Darwinion. Buenos
Aires, etc.

Dasika chronika. Athenai

Debreceni Agrártudományi Főiskola
Tudományos Közleményei Debrecen

Debreceni Agrártudományi Főiskola
Tudományos Ulésszakának
Előadásai. Debrecen

Decheniana. Verhandlungen des
Naturhistorischen Vereins der
Rheinlande und Westfalens.
Bonn

Deciduous Fruit Grower. Cape Town

Deer. Journal of the British
Deer Society. Southampton

Defence Science Journal. New Delhi

Défense des végétaux. Paris

Deghekkakir, Haykakan SSR
Kidutyunneri Akademia:
Biolokiakan Kidutyunner

Delphinium Society's Year Book.
London

Delta del Paraná. Estación
Experimental Agropecuaria Delta
del Paraná. Campana, Argentina

Deltion tēs 'Ellēnikēs geōgrafikēs
'etaireias. Athenais.

Demografia. Budapest

Den'gi i Kredit, Moskva.

Department of Agricultural
  Economics Research Report.
  Louisiana Agricultural Experi-
  ment Station, Baton Rouge, La.

Departmental Memorandum.
  Federal Department of Agricultural Research, Ibadan

Departmental Publication Forestry
  Branch. Department of Forestry
  and Rural Development. Canada

Departmental Report. Director of
  Agriculture, Fisheries and
  Forestry, Hong Kong

Derevoobrabatÿvayushchaya
  promÿshlennost'. Moskva

Dermatologia. Mexico

Dermatologia venezolana. Caracas

Dermatologica. Basel

Dermatologische Monatschrift.
  Leipzig

Dermato-venerologia. Bucureşti

Desarrollo economico. Buenos
  Aires

Deutsche Agrartechnik. Berlin

Deutsche Aussenpolitik. Berlin

Deutsche Baumschule. Aachen

Deutsche entomologische
  Zeitschrift. Berlin

Deutsche Gartenbau. Berlin

Deutsche Gartenbauwirtschaft.

Deutsche Gartnerborse

Deutsche Gesundheitswesen. Berlin

Deutsche Lebensmittel-Rundschau.
  Nürnberg, etc.

Deutsche medizinische Wochenschrift. Leipzig, etc.

Deutsche Milchwirtschaft
  (Molkerei- und Käserei-Zeitung).
  Hildesheim

Deutsche Molkerei-Zeitung.
  Kempten

Deutsche Pilztierzüchter. München

Deutsche Schlachthofzeitung.
  München

Deutsche tierärztliche
  Wochenschrift. Hanover

Deutsche tropenlandwirt,
  Witzenhausen a.d. Werra

Deutsche Weinbau. Berlin

Deutsche Wirtschaftsgeflugelzucht.

Deutsches Wirtschaftsinstitut
  Berichte. Berlin

Deutsche Zeitschrift fur Nerven-
  heilkunde. Leipzig

Deutsches medizinisches Journal.
  Berlin

Developing economies. Tokyo

Development Studies. Department
  of Agricultural Economics,
  University of Reading, Reading

Developments in Industrial
  Microbiology. New York

Developmental Biology. New York,
  London

Developpement et Civilisations,
  Paris

Diabetes. New York

Diabetologia. Berlin

Día médico. Buenos Aires

Die grune. Zurich

Dietologia. Buenos Aires

Dirigente rural. São Paulo

'Discovery' Reports. Cambridge

Discussion Paper. Department of
  Agricultural Economics and
  Farm Management, Massey Univer-
  sity, Palmerston North. NZ

Discussions of the Faraday
  Society. London

Diseases of the Chest. El Paso,
  Texas, etc.

Diseases of the Colon and Rectum.
  Philadelphia

Diseases of the Nervous System.
  Chicago

Dissertation Abstracts. Ann Arbor

Dissertation. Landwirtschaftliche
  Fakultät der Justus Leibig-
  Universität. Giessen

Distribution Maps of Pests (Series A) CIE. London

Distribution Maps of Plant Diseases. CMI. London

District Bank Review. Manchester

Divisional Report. Division of Plant Industry. CSIRO. Australia Canberra

Divisional Report. Division of Soils. CSIRO Australia. Adelaide

Divisional Report. Land Research Regional Survey. CSIRO Australia. Melbourne

Divulgacao agronomica. Piracicaba

Document. International Poplar Commission. FAO

Documenta de medicina geographica et tropica. Amsterdam

Documenta Veterinaria. Czechoslovakia

Documenta Veterinaria Publicationes Instituti Medicinae Veterinariae Brunensis. Prague

Documentation in Agriculture and Food. Organization for Economic Co-operation and Development, Paris

Documentation Bulletin. National Research Centre. Cairo

Documentation sur l'Europe Centrale Louvain

Documents Techniques. Institut National de la Recherche Agronomique de Tunisie, Ariana

Documents. Timber Committee, Economic Commission for Europe

Doklady Akademii nauk Armyanshoi SSR. Erevan

Doklady Akademii nauk Azerbaidzhanskoi SSR. Baku

Doklady Akademii nauk BSSR

Doklady Akademii nauk Belorusskoi SSR. Minsk

Doklady Akademii nauk SSSR. Moskva, Leningrad

Doklady Akademii-nauk Tadzhikskoi SSR. Stalinabad

Doklady Akademii nauk Uzbekskoi SSR. Tashkent

Doklady Akademii Sel-sko-khozyaistvennykh Nauk i Bolgarii Sofia

Doklady Botanical Sciences (USSR Acad. Sci.) New York

Doklady Instituta Geografii Sibirskogo Dal'nogo Vostoka

Doklady k 5-mu Mezhdunarodnomu Kongressu po Mekhanike Gruntov i Fundamentostroeniya, Akademiya Stroitel'stva i Arkhitektury SSSR. Gosstroiiadat

Doklady Moskovskoi sel'sko-khozyaist-vennoi akademii im. K.A. Timiryazeva. Moskva

Doklady Rossiiskoi Sel'skoi-hozyaistvennoi Akademii im. K.A. Timiryazeva. Moskva.

Doklady Sibirskikh Pochvovedov, Novosibirska. Sibirskoe Otdele-nie Akademii nauk. Novosibirsk

Doklady Soil Science. Madison

Doklady Soobšćenija VNIIESH. Moskva

Doklady Vsesoyuznoi akademii sel'skokhozyaistvennykh nauk im V.I. Lenina

Dopovidi Akademiyi nauk Ukrayins'koyi RSR. Kiev

Doriana. Supplemento agli Annali del Museo civico di Storia naturale "Giacomo Doria", Genoa

Dřevársky výskum. Bratislava

Dřevo. Praha

Droit social. Paris

Drvna industrija. Zagreb

Duna-Tisza Közi Mezőgazdasági Kisérleti Intézet Bulletinje. Kecskemet.

Dunantuli Tudomanyos Gyujtemeny.

Duodecim. Helsinki

Dupont Agricultural News Letter

Durban Museum novitates. Durban

E

Earth-Science Reviews. Amsterdam

East African Agricultural and Forestry Journal. Nairobi

East African Journal of Rural Development. Nairobi

East African Medical Journal. Nairobi

East African Wildlife Journal. Kenya

Eastern European Economics. White Plains, New York

Eastern Horizon. Hong Kong

Écho médical du Nord. Lille

Ecole National Veterinaire d'Alfort. Theses. Alfort

Ecological Monographs. Durham, NC

Ecological Review. Mount Hokkada Botanical Laboratory, Sendai, etc

Ecology. Brooklyn, etc.

"Ecology of Soil-borne Pathogens". Proceedings of an International Symposium on Factors determining the behaviour of Plant Pathogens in Soil, held at the University of California, April 7-13 1963. London

Econometric Annual of the Indian Economic Journal. Bombay

Econometrica. Amsterdam

De Economia. Madrid

Economia y agricultura. Lima

Economia internazionale. Geneva

Economic Affairs. Calcutta

Economic Annalist. Department of Agriculture, Canada. Ottawa

Economic Botany. N.Y. Botanical Garden. Lancaster, Pa.

Economic Bulletin for Africa. Addis Ababa

Economic Bulletin for Asia and the Far East. New York

Economic Bulletin, Bank of Libya. Tripoli

Economic Bulletin of Ghana. Accra

Economic Bulletin, Hawaii Experiment Station, Honolulu.

Economic Bulletin for Latin America. New York

Economic Department Report. West of Scotland Agricultural College, Glasgow

Economic Development and Cultural Change. Chicago

Economic Development Report. The Centre for International Affairs, Harvard University, Cambridge. Mass.

Economic Geography. Worcester, Mass. etc.

Economic Geology and Bulletin of the Society of Economic Geologists. Lancaster, Pa. etc.

Economic Information Report. Department of Agricultural Economics, University of Minnesota, St. Paul, Minn.

Economic Information Report. Department of Economics, North Carolina State University, Raleigh, N.C.

Economic Journal. London

Economic Planning, Montreal

Economic and Political Weekly. Bombay

Economic Proceedings of the Royal Dublin Society. Dublin

Economic Quarterly. Tel Aviv

Economic Report. Agricultural Economics Division, North of Scotland College of Agriculture. Aberdeen

Economic Report. North of Scotland College of Agriculture. Aberdeen

Economic Report. US Department of Agriculture. Washington

Economic Research Report. Department of Economics, North Carolina State University. Raleigh, NC

Economic Review. Helsinki

Economic Review, Bank of China.
Taipei

Economic Series. British Museum
Natural History. London

Economic Series. Department of
Agricultural Economics and
Marketing. Pretoria

Economic Study Report. Department
of Agricultural Economics,
University of Minnesota. St.
Paul. Minn.

Economic Survey. Buenos Aires

Economic Weekly. Bombay

Economica. London

Economie Appliquée. Paris

Economie et Humanisme. Caluire
(Rhône)

Economie et médecine animales.
Paris

l'Economie neridionale.
Montpellier

Economie et politique. Paris

Economie rurale (agrotike
oikonomia). Athens

Economie rurale. Beirut

Economie rurale. Paris

Economie et Statistique. Paris

Economische statistische
berichten. Rotterdam

Economist. Haarlem

Economist. London

Edinburgh Medical Journal.
Edinburgh

Eesti loodus. Tallinn

Eesti NSV teaduste akadeemia
toimetised. Tallinn.

Bioloogiline seer

Egészségtudomány. Budapest

Egypte contemporaine. Cairo

Egyptian Agricultural Review.
Cairo

Ehime Daigaku Nogakubu Kiyo.
Matsuyama

Einheit. Berlin

Eiszeitalter und Gegenwart.
Hannover

Ek Ton Ergazterion Edafologias
Ton Aristoteleion Panespistemion
Thessalonikis. Thessalonikh.

Ekistics. Athens

Ekologia polska. Warszawa.

Ser. A.

Ekonomi Forskningstiftelsen
Skogsarbeten. Stockholm

Ekonomiceskie nauki. Moscow

Ekonomika i matematiceskie Métody.
Moskva

Ekonomika i organizacja rolnictwa.
Warszawa

Ekonomika poljoprivrede. Beograd

Ekonomika sel'skogo khozyaistva.
Moskva

Ekonomika zemedelstvi. Praha

Ekonomicky casopis. Bratislava

Ekonomiak Revy. Stockholm

Ekonomista. Warszawa

Ekonomski pregled. Zagreb

Eksperimental'naya Botanika.
Institut Biologii, Akademiya
Nauk Belorusskoi. SSR

Elanco Agricultural Chemicals
Newsletter. London

Electrical Research Association
Digest.

Electroencephalography and Clinical
Neurophysiology. Montreal

Électronique industrelle. Paris

Élelmezési ipar. Budapest

Élemiszertudomany. Budapest

Élelmiszervissgalati közlemenyek.
Budapest

Élevage et insemination. Paris

Ellēnikē iatrikē. Thessalonikē.

Ellenike Kteniatrike. Thessaloniki

Emballages. Paris

Empire Forestry Review. London

Encyclopedia of Plant Physiology. Berlin

Encyclopédie entomologique. Paris

Endeavour. London

Endemic Diseases Bulletin. Nagasaki University. Nagasaki

Endocrinologia japonica. Tokyo

Endocrinology. Glendale, Cal.etc.

Endokrinologie. Leipzig

Energia nucleare.in agricoltura. Milano

Engineering Geology. Association of Engineering Geologists. Sacramento

Entomologia experimentalis et applicata. Amsterdam

Entomological Circular. Department of Agriculture, British Columbia. Victoria, B.C.

Entomological News. Academy of Natural Sciences, Philadelphia

Entomological Review. Washington

Entomological Review of Japan. Osaka

Éntomologicheskoe obozrenie. Moskva.

Entomologie et phytopathologie appliquées Tehran

Entomologische abhandlungen und berichten aus dem staatlichen museum fur tierkunde in Dresden. Dresden

Entomologische Arbeiten aus dem Museum Georg Frey. München

Entomologische berichten. Nederlandsche entomologische vereeniging. Amsterdam

Entomologische blatter. Krefeld

Entomologische Mitteilungen aus dem Zoologischen Staatsinstitut und Zoologischen Museum Hamburg. Hamburg

Entomologisk tidskrift. Stockholm

Entomologiske Meddelelser. Kjøbenhavn

Entomologiste. Paris

Entomologist's Gazette. Feltham

Entomologist's Monthly Magazine. Oxford

Entomology Circular. Florida Department of Agriculture, Division of Plant Industry. Gainesville

Entomology Memoirs. Department of Agricultural Technical Services. Republic of South Africa. Pretoria

Entomology Newsletter. Department of Agriculture, Canada. Ottawa

Entomophaga. Paris

Entreprise Agricole. Paris

Environmental health. India

Enzymologia: Acta biocatalytica. Den Haag.

Enzymologia biologica et clinica. Basle

Eos. Revista española de entomología. Madrid

Epidemiological and Vital Statistics Report. WHO. Geneva

Epistèmonike epetèris, Geoponike kai dasologike schole, Aristoteleion panepistemion Thessalonikes.

l'Equipment agricole. Paris

Equipment Evelopment and Test Report. US Forest Service

Erde. Berlin

Erdészeti és Faipari Eypetem Kiadványai, Sopron

Erdészeti as Faipari Egyetem Tudományos Közleményei. Sopron

Erdészeti kutatások. Budapest

Erdészettudományi közlemények. Sopron

Erdkunde. Bonn

Erdő. Budapest

Ergebnisse der Allgemeinen Pathologie und Pathologischen Anatomie

Ergebnisse der Biologie. Berlin

Ergebnisse der inneren Medizin und
Kinderheilkunde. Berlin

Ergebnisse der landwirtschaftlichen Forschung an der Justus
Leibig-Universität. Giessen

Ergebnisse der wissenschaftlichen Untersuchung des
Schweizerischen Nationalparks.

Erhvervsfrugtavleren. Odense

Ernährungsdienst. Deutsche
Getreidezeitung. Hannover

Ernährungsforschung. Berlin

Ernährungs-Umschau. Frankfurt am
Main

Erwerbobstbau. Berlin

Esakia. Occasional Papers of the
Hikosan Biological Laboratory
in Entomology, Hikosan

Essex Farmers' Journal. Chelmsford

Esso Farmer. London

Estanzuela. Uruguay

Estate Gazette. London

Estudios Sindicales y
Cooperativos. Madrid

Estudos agronomicos. Lisboa

Estudos e informação. (Direcção
geral dos) servicos florestais
e aqüicolas. Lisboa

Etesion Deltion. Upourgeion
Georgias. Futopathologikos
Stathmos Patron. Report.
Ministry of Agriculture
Phytopathological Station.
Patras

Ethnology. Pittsburg

Etizenia. Fukui, Japan

Etlik veteriner bakterioyoloji
enstitüsü dergisi. Ankara

Études du Centre National d'Études
et d'Experimentation de
Machinisme Agricole, Antony-Seine

Études et conjoncture. Paris

Études d'economie rurale. Rennes

Études Nigériennes, Centre
National de la Recherche
Scientifique. Paris

Etudes rurales. Paris

Etudes senegalaises. Dakar

Etudes statistiques. Paris

Euphytica. Wageningen

Europe France Outre Mer. Paris

European Journal of Biochemistry.
Berlin

European Journal of Pharmacology.
Amsterdam

European Packaging Digest. Paris

European Potato Journal.
Wageningen

Evolution. USA

Excerpta botanica. Stuttgart

Sectio A. Taxonomica et
chlorologica

Excerpta medica. Amsterdam

Sectio 3. Endocrinology

Excerpta Veterinaria Lublin.
Poland

Experientia. Basel

Experientiae.

Experiment Station Circular.
South Dakota Agricultural
Experiment Station, Brookings.

Experimental Agriculture. UK

Experimental Brain Research.
Berlin

Experimental Cell Research.
New York

Experimental Horticulture.
London

Experimental Husbandry. London

Experimental Medicine and
Surgery. New York, etc.

Experimental and Molecular
Pathology. New York

Experimental and Molecular
Pathology. Supplements. New York

Experimental Parasitology.
New York

"Experimental Pedology". Proceedings
of the Eleventh Easter School in
Agricultural Science, University
of Nottingham, 1964. London

Experimental Record of the
Department of Agriculture,
South Australia

Experimental Reports of Equine
Health Laboratory. Japan

Experimental Report of Government
Experimental Station for Animal
Hygiene. Tokyo

Experimental Work. Edinburgh
School of Agriculture.

Experimentelle Veterinärmedizin.
Leipzig

Exploration du Parc National
Albert. Deuxième Série.
Bruxelles

Exploration du Parc National de
la Garamba. Mission H. de
Saeger. Bruxelles

Exploration du Parc National de
l'Upemba. Mission G.F. de
Witte. Bruxelles

Extension en las Americas

Extension Bulletin. Cornell
Agricultural Experiment Station.
Ithaca, N.Y.

Extension Bulletin Farm Science
Series. Cooperative Extension
Service, Michigan State
University, East Lansing, Mich.

Extension Bulletin. Hawaii
College of Agriculture

Extension Bulletin. New Jersey
College of Agriculture.
New Brunswick

Extension Bulletin. Washington
State College

Extension Circular. College of
Agriculture, University of
Illinois, Urbana, Ill.

Extension Circular. Hawaii
College of Agriculture

Extension Circular. North
Carolina Agricultural Extension
Service

Extension Circular. North
Carolina State College of
Agriculture and Engineering

Extension Folder. Agricultural
Extension Service, North
Carolina State College of
Agriculture and Engineering

F

F.A.O. Agricultural Studies

F.A.O. Atomic Energy Series.
Washington

F.A.O. Development Paper.
Agriculture

F.A.O. Forestry and Forest Product
Studies. Roma

F.A.O. Plant Protection Bulletin.
Rome

F.A.O. Report. Rome

F.A.O. Soils Bulletin. Rome

F.A.O. World Consultation on Forest
Genetics and Tree Improvement.
Stockholm, Rome.

F.I.R.A. Bulletin

Fachliche Mitteilungen der
österreichischen Tabakregie.
Wien

Faipari Kutatások. Budapest

Far Eastern Economic Review.
Hong Kong

Farm Building Progress.
Aberdeen

Farm Buildings Digest. Kenilworth.

Farm Bulletin. Department of
Agriculture, Irish Republic.
Dublin

Farm Bulletin (New Series). Indian
Council of Agricultural Research
New Delhi

Farm Chemicals. Philadelphia

Farm and Country. London

Farm Economics. Ithaca, NY

Farm Economist. Oxford

Farm and Factory. Madras

Farm Forestry. Wellington, N.Z.

Farm and Home Science. Logan

Farm Implement and Machinery
Review. London

Farm Incomes, Costs and
Management

Farm Management. Kenilworth

Farm Management. Melbourne

Farm Management Notes for Asia
and the Far East. Bangkok

Farm Management Notes. University
of Nottingham, School of
Agriculture. Sutton Bonington.

Farm Management Quarterly.
Aberdeen

Farm Management Report. Faculty
of Agricultural Economics,
University of New England,
Armidale

Farm Management Survey Report.
Western Research Centre,
Ministry of Agriculture, Mwanza

Farm Mechanization and Buildings.
London

Farm Policy. Perth

Farm Policy Forum. Ames, Iowa

Farm Quarterly. Cincinnati

Farm Research. Geneva, N.Y.

Farm Research News. Dublin

Farmaco. Pavia

Farmacognosia. Madrid

Farmakologiya i Toksikologiya.
Moskva

Farmbuildings

Farmer and Forester. Zombia

Farmer. Journal of the Jamaica
Agricultural Society. Kingston

Farmer and Stock Breeder. London

Farmers' Bulletin. US Department
of Agriculture, Washington

Farmers' Leaflet. National
Institute of Agricultural
Botany. Cambridge

Farmers' Newsletter. Lecton, NSW

Farmers Report, Department of
Agricultural Economics,
University of Leeds. Leeds

Farmers Report. Economic Division
School of Agricultural Science,
University of Leeds. Leeds

Farmer's Weekly. Bloemfontein

Farmers' Weekly. London
Farming. Norwich

Farming Progress (Pertwee)

Farming Review. Edinburgh

Farming in South Africa. Pretoria

Farming World. Norwich

Farskotsel. Svenska Faravels-
foreningens Tidskrift. Sweden

Fauna. Norsk zoologisk forenings
tidskrift. Oslo

Fauna and Flora. South Africa

Fauna och flora. Uppsala

Fauna SSSR. Leningrad

Faune de France. Paris

Faunistische Abhandlungen
Staatliches Museum fur Tierkunde
in Dresden. Dresden

FEBS Letters. Federation of
European Biochemical Societies.
Amsterdam

Feddes Repertorium specierum
novarum regni vegetablis

Federal Register. Washington

Federal Veterinarian. Kansas City

Federation Proceedings.
Federation of American Societies
for Experimental Biology.
Baltimore

Feed Forum. UK

Feedstuffs. Minneapolis

Feld und Wald. Essen

Fel'dsher i akusherka. Moskva

Felsöoktatási Szemle. Budapest

Fertiliser Feed and Pesticide
Journal. London

Fertiliser and Feeding Stuffs
Journal. London

Fertiliser News. New Delhi

Fertilité. Information on tropical and subtropical fertilisation
Paris

Fertility and Sterility.
New York

Fertilmacchine. Bologna
Fertodi novenynemesitesi es
novenytermesztesi kutato intezet
kozlemenyei

Fette und Seifen (einschliesslich
der Anstrichmittel). Berlin,
etc.

Feuille des naturalistes. Paris

Fibra. Wageningen

Field. London

Field Crop Abstracts. Aberystwyth,
etc.

Field and Laboratory. Contributions
from the Science Departments of
Southern Methodist University,
Dallas

Field Station Record, Division of
Plant Industry, CSIRO

Field Studies. London

Fieldiana: Botany. Chicago

Figyelö. Budapest

Fiji Agricultural Journal. Suva

Fiji Farmer. Suva

Fiji Timbers and their uses.
Department of Forestry, Suva

Finance and Development.
Washington

Finance a uver. Praha

Financial Report. North of
Scotland College of Agriculture,
Aberdeen

Finanse. Warszawa

Finansi i Kredit. Sofija

Finansy SSSR. Moskva

Finnish Paper and Timber. Helsinki

Finsk Veterinartidskrift. Finland

Fire Control Note, California
Division of Forestry, Sacramento

Fire Control Notes. US Forest
Service. Washington

Fire Protection Yearbook. London

Fire Research Abstracts and
Reviews. Washington

Fishery Bulletin. Fish and Wildlife Service. US Department of
Interior. Washington

Fishery Leaflet. Fish and Wildlife Service. US Department of
Interior. Washington

Fishery Research and Management
Division Bulletin. Department of
Inland Fisheries and Game.
State of Maine. Augusta

Fishing News International.
London

Fiskeridirektoratets skrifter.
Bergen.

Serie Teknologiske undersøkelser

Fisons Agricultural Technical
Information

Fitofilo. Direccion general de
agricultura. San Jacinto, Mexico

Fitopatología. Bogotá

Fitopatologia. Santiago

Fitosanitarias. Organo del
Departamento de Sanidad Vegetal
de la Facultad de Agronomia de
la Universidad de la Plata.
La Plata

Fitotecnia latinoamericana.
Costa Rica

Fiziologicheskii zhurnal SSSR im
I.M. Sechenova. Leningrad

Fiziologichnie-Biokhimichnie
Osnovi Pidvishchennya Produktiv-
nesti Poslin. Kiev.

Fiziologiya rastenii. Moskva

Fleischwirtschaft. Frankfurt-am-
Main

Flora, oder allgemeine botanische
Zeitung. Abt. A. Physiologie
und Biochemie. Abt. B. Morphologie und Geobotanik. Jena.
Regnesburg

Flora og Fauna. Silkeborg.

Florida Entomologist. Gainesville

Florida Grower

Florist and Nursery Exchange

Flugblatt. Biologische
Bundesanstalt für Land- und
Forstwirtschaft. Berlin

Flugblatt. Bundesanstalt für
Pflanzenschutz. Wien

Flugblatt der Eidge Landw. Versuch-
sanstalt Zurich-Oerlikon

Flugblatt. Verein für Zucherrüben-
forschung. Haringsee

Flygblad. Statens växtskyddsanstalt
Stockholm

Foderation europaeischer
gewasserschutz informationsblatt.

Földrajzi értesítő. Budapest

Földrajzi közlemények. Kiadja a
Magyar fo"ldrajzi tarsaság.
Budapest

Folha medica. Rio de Janeiro

Folia biologica. Buenos Aires

Folia biologica. Krakow

Folia biologica. Praha

Folia clinica et biologica. Brazil

Folia endocrinol'ogica. Pisa

Folia endocrinologica japonica.
Kyoto

Folia entomologica hungarica.
Budapest

Folia Forestalia Polonica. Komitet
nauk lesnych, wydzial nauk
rolniczych i lesnych

Seria A. Lesnictwo

Folia Forestalia Instituti
Forestalis Fenniae. Helsinki

Folia geobotanica et phyto-
toxonomica

Folia geobotanica and phyto-
taxonomica bohemoslovaca.
Czechoslovak Academy of
Sciences. Prague

Folia medica. Cracoviensia

Folia medica. Napoli

Folia medica. Plovdiv

Folia microbiologica. Praha
Folia Parasitologica.
Czechoslovakia

Folia psychiatrica, neurologica
et neurochirurgica neerlandica.
Amsterdam

Folia veterinaria. Facultatis
Medicinae veterinariae cassovi-
ensis. Kosice, Czechoslovakia

Folleto de divulgación. Secretaria
de agricultura y ganaderia,
México

Folleto miscelaneo. Secretaria de
agricultura y ganaderia,
Mexico

Folleto tecnico. Secretaria de
agricultura y ganaderia,
Mexico

Folletos tecnicos forestales,
Administracion Nacional de
Bosques, Buenos Aires

Fomento. Técnica s Economia
Ultramarinas. Lisboa

Food Additives Control Series.
FAO. Rome

Food and Agricultural Legislation
FAO. Rome

Food in Brief (including Cheese
Abstracts). New York

Food and Cosmetics Toxicology. UK

Food Engineering. New York

Food Industries of South Africa.
Cape Town

Food Industry Studies. United
Nations, New York

Food Irradiation. Saclay

Food Manufacture. London

Food and Nutrition Notes and
Reviews. Canberra

Food Preservation Quarterly.
Sydney

Food Processing. Chicago (Formerly
Mgnt. Fd. Process./Mktg.)

Food Processing Industry. London
(formerly Fd.Process.Mktg.)

Food Research Institute Studies.
Stanford University, Stanford.
Food Science and Technology
Abstracts. IFIS

Food Technology. Champaign, etc.

Food Technology in Australia. Sydney

Forage Notes. Ottawa

Forderungsdienst. Austria

Foreign Affairs. New York

Foreign Agricultural Economic Report. Economic Research Service US Department of Agriculture, Washington

Foreign Agricultural Economic Report. US Department of Agriculture, Washington

Foreign Agricultural Trade of the United States. Washington

Foreign Agriculture. US Department of Agriculture. Washington

Foreign Agriculture Circular. US Department of Agriculture, Foreign Agriculture Service. Washington

Foreign Agriculture Report. US Department of Agriculture. Washington

Foreign Trade Review. New Delhi

Foreningen Skogtradsforadling Arsbok

Foreningen for Vaxtforadling av Frukttrad

Forest Department Bulletin, Forest Department, Ndola, Zambia

Forest Economy. Tokyo

Forest Division Technical Note. Forest Division, Ministry of Agriculture, Forests and Wildlife. Dar-es-Salaam

Forest Fire Control Abstracts. Department of Forestry and Rural Development, Ottawa

Forest Industries. Portland

Forest Management Notes. Forest Service, Department of Lands and Forests. British Columbia Forest Service. Victoria

Forest Pest Leaflet. Forest Service. Washington. DC

Forest Products Journal. Madison

Forest Products News Letter. CSIRO. Melbourne

Forest Products Research. Department of Scientific and Industrial Research. London

Forest Products Research Records Department of Scientific and Industrial Research. London

Forest Products Research Reports, Department of Forest Research, Ibadan, Nigeria

Forest Record. Forestry Commission London

Forest Research Bulletin. Forest Department. Zambia. N'dola

Forest Research Note. College of Agriculture, Wisconsin University. Madison

Forest Research Pamphlet, Division of Forest Research. Kitwe, Zambia

Forest Research Review. Forest Service, British Columbia. Victoria

Forest Research Review. Forestry Division, Department of Lands. Dublin

Forest Resources Newsletter. Canberra

Forest Resource Report. US Forest Service. Washington

Forest Science. Washington

Forest Science Monographs. Society of American Foresters. Washington

Forest-Soil Relationships in North America. Second Northern American Soils Conference. Corvallis. Oregon

Forest Survey Notes. British Columbia Forest Service. Victoria

Forest and Timber. Sydney

Forester. Northern Ireland

Forestry. London, etc.

Forestry Abstracts. Oxford

Forestry Abstracts Leading
   Article Series, Commonwealth
   Forestry Bureau, Oxford

Forestry Chronicle, Macdonald
   College, Quebec

Forestry Chronicle. Toronto

Forestry Commission Leaflet. London

Forestry Development Paper. FAO

Forestry Economics. University of
   the State of New York. College
   of Forestry. Syracuse

Forestry Equipment Notes.
   FAO Rome

Forestry Note. Illinois University
   Agricultural Experiment Station.
   Urbana

Forestry Occasional Paper.
   FAO Rome

Forestry Research Highlights.
   Report Rocky Mountain Forest
   and Range Experiment Station.
   Fort Collins

Forestry Research Notes.
   University of Wisconsin College
   of Agriculture. Madison

Forestry in South Africa.
   Pretoria

Forestry Technical Note. EAAFRO

Forestry Technical Notes.
   University of New Hampshire
   Agricultural Experiment Station.
   Durham

Forestry Technical Papers.
   Forestry Commission of Victoria.

La Forêt: Organe de la Société
   Forestière Suisse et de
   l'Association Suisse d'Économie
   Forestiere, Neuchâtel

Forêt privee. Paris

Formosan Science. Formosan
   Association for the Advancement
   of Science. Taipei

Forschung und Beratung, Forstwirtschaft. Düsseldorf

Forschungen und Fortschritte.
   Korrespondenzblatt (Nachrichtenblatt) der deutschen Wissenschaft und Technik. Berlin

Forskning og forsøk i landbruket.
   Oslo

Forschungen auf dem Gebiet der
   Pflanzenkrankheiten. Tokyo

Forskningsresultater fra Lands-
   foreningen for Kosthold og
   Helse. Oslo

Forsøg och forskning. Stockholm

Forstarchiv. Hannover

Forst-und Holzwirt. Hannover

Forstlige Forsoksvaesen i Danmark
   kjøbenhavn

Forsttechnische Informationen.
   Mainz.

Forstwirtschaftliches Zentralblatt
   Hamburg

Forstwissenschaftliche Forschung:
   Beihefte zum Forstwissenschaft-
   lichen Zentralblatt. Hamburg

Forstwissenschaftliches Zentralblatt. Berlin

Fort Hare Papers. Fort Hare
   South Africa

Fortpflanzung, Zuchthygiene und
   Haustierbesamung. Hannover

Fortschritte auf dem Gebiete der
   Röntgenstrahlen. Hamburg, etc.

Fortschritte in der Geologie von
   Rheinland und Westfalen.

Fortschritte der Neurologie und
   Psychiatrie und ihrer Grenzgebiete. Leipzig, Stuttgart

Fortschrittsbericht für die
   Landwirtschaft. Berlin

Fortschrittsberichte für Landwirtschaft und Nährungsgüterwirtschaf
   Berlin

Forum. Geneva

Fourrages. France

Fragmenta balcanica Musei
   macedonici scientiarum
   naturalium. Skopje

Fragmenta faunistica. Warszawa

Fragmenta floristica et geobotanica
   Krakow.

France médicale. Paris

Frankfurter Zeitschrift fur
Pathologie. Wiesbaden

Freedom from Hunger Campaign
Basic Study, Food and Agricultural Organization. Rome

Freedom from Hunger Campaign
Report. FAO Rome

Freiberger Forschungshefte.
Bergakademie. Freiberg.

Freyr. Mánaðarrit um landbúnað.
Reykjavik

Friesia. Nordisk mykologisk
Tidsskrift. Kóbenhavn

Friuli medico. Udine

Frontiers. A magazine of natural
history. Philadelphia

Frontiers of Plant Sciences.
New Haven

Fruit Annual

Fruit belge. Gembloux

Fruit Grower. London

Fruit and Produce. Auckland

Fruit Varieties and Horticultural
Digest. Wooster, Ohia, etc.

Fruit and Vegetable Crop Prospects
OEEC. Paris

Fruit World and Market Grower

Fruit Year Book. Horticultural
Society. London

Fruits. Paris

Fruits d'outre mer. Paris

Fruits et primeurs de l'Afrique
du Nord. Casablanca

Fruitteelt. Arnhem

Frukt i ar. Stockholm

Frukt og baer. Norsk hageselskap.
Oslo

Fruktodlaren. Stockholm

Frustula entomologica. Fano

Frutos

Frutticoltura. Bologna

Fuji Bank Bulletin. Tokyo

Fukuoka acta medica. Fukuoka

Fukushima Journal of Medical
Sciences. Fukushima

Fur Trade Journal of Canada.
Oshawa, Ont.

G

G.E.N. Sociedad venezolana de
gastroenterologia, endocrinologia
y nutrición. Caracas.

G.F.M.-Mitteilungen zur
Markt- und Absatzforschung.
Hamburg

Gaceta médica. Guayaquil

Gaceta médica de Caracas. Caracas

Gaceta médica de Mexico. Mexico

Gaceta veterinaria. Buenos Aires

Ganaderia. Revista del sindicato
nacional de Ganaderia. Madrid

Gann (Ergebnisse der Krebsforschung in Japan). Tokyo

Garcia de orta. Revista da Junta
das missões geográficas e de
investigacões do ultramar.
Lisboa

Garden Journal of the New York
Botanical Garden. New York

Gardeners' Chronicle and Gardening
Illustrated. London

Gardening Illustrated. London

Gardens' Bulletin. Singapore

Gartenbauwirtschaft. Germany

Gartenbauwissenschaft. Berlin,
München

Gartenwelt. Hamburg

Gartner Tidende. Denmark

Gartneryrket

Gastroenterologia. Basel

Gastroenterologia bohema. Praha

Gastroenterology. Baltimore

Gazdalkodas. Budapest

Gazdaság. Budapest

Gazeta Agricola de Angola. Luanda

Gazeta do agricultor, Moçambique.
Lourenço Marques.

Gazeta cukrownicza. Warszawa

Gazeta médica portuguesa. Lisboa

Gazette medicale de France. Paris

Gazzetta internazionale di
  medicina e chirurgia. Rome
Gazzetta medica italiana. Turin

Gazzetta veterinaria. Milano

Geburtshilfe und Frauenheilkunde.
  Stuttgart

Geflügelhof.

Genenbaurs morphologisches
  jahrbuch. Germany

Genen en phaenen. Wageningen

General and comparative endocrinology. UK

General Report. Farmer Cooperative
  Service. US Department of
  Agriculture, Washington

Genetica. 's Gravenhage

Genetica agraria. Roma, etc.

Genetica iberica. Madrid

Genetica polonica. Poznan

Genetical Research. Cambridge, etc.

Genetics. Princeton, Austin, etc.

Genetika. USSR

Genetika a slechteni

Genéve-Afrique. Carouge-Genéve

Génie rural. Paris

Genio rurale. Bologna, etc.

Gentes herbarum. Ithaca. N.Y.

Geochimica et cosmochimica acta.
  London

Geofisica e meteorologia. Genova

Geograficky časopis. Praha

Geograficheskii sbornik. l'vovogo
  instituta. l'vovogo otdela
  geograficheskogo obshchestva
  SSSR. Moskva

Geografiska Annaler. Stockholm

Geographical Abstracts. London

Geographical Journal. Royal
  Geographical Society. London

Geographical Magazine. London

Geographical Review. New York, etc.

Geographical Review of India.
  Calcutta

Geographische Berichte. Berlin
Geographische Rundschau.
  Braunschweig

Geographische Zeitschrift. Leipzig,
  Wiesbaden

Geography. London

Geologica bavarica. München

Geologichnii zhurnal. Akademiya
  Nauk Ukrainskoi SSR. Kiev

Geologie. Berlin

Geologisches Jahrbuch. Amt für
  Bodenforschung. Hannover

Geomorphological Abstracts. London

Geoponika. Periodos B. Greece

Georgia Agricultural Research.
  Athens

Georgia Veterinarian

Georgikon deltion. Greece

Georgofili. Atti della Accademia
  dei georgofili. Firenze

Geotechnique. London

Geriatrics. Minneapolis

German Economic Review. Stuttgart

Gesnerus. Vierteljahrsschrift für
  Geschichte der Medizin und der
  Naturwissenschaften. Zürich

Gestencilde Verslagen van
  Interprovinciale Proeven.
  Proefstation voor de Akker- en
  Weidebouw. Wageningen

Gestion. Paris

Gesunde Pflanzen. Frankfurt,
  Stuttgart

Gesundheitsingenieur. Berlin

Getreide, Mehl und Brot

Getreide und Mehl (vereinigt mit
  Muhlenlaboratorium). Detmold

Gewässer und Abwässer. Düsseldorf

Gewerkschaftliche Monatshefte.
  Köln

Ghana Farmer. Accra

Ghana Journal of Agricultural
Science. Accra

Ghana Journal of Science

Giannini Foundation Monograph.
California Agricultural
Experiment Station, Berkeley

Giannini Foundation Research Report
California Agricultural
Experiment Station, Berkeley

Gidroliznaya i lesokhimisheskaya
promyshlennost'. Moskva

Gidrometeorologiya Azerbaidzhana
i Kaspilskogo. Morya, Baku

Gidrotekhnika i melioratsiya.
Moskva

Giessener abhandlungen zur agrar-
und wirtschaftsforschung des
europaischen ostens. Giessen

Giessener schriftenreih tierzucht
und haustiergenetik. Giessen

Gigiena i sanitariya. Leningrad,
Moskva

Giornale di batteriologia
viroligia e immunologia.
Torino

Giornale botanico italiano.
Firenze

Giornale digli economisti e annali
de economia. Padova

Giornale di igiene e medicina
preventiva. Italy

Giornale italiano di chirurgia.
Napoli

Giornale italiano di dermatologia
(formerly e sifilografia).
Milano

Giornale de malattie infettive e
parassitarie. Italy

Giornale di medicina militare.
Roma

Giornale di microbiologia. Milan

Giornale di scienze mediche.
Venezia

Gladiolus. Boston, Mass

Glas Srpske akademije nauka.
Beograd

Glasgow Medical Journal. Glasgow

Glasgow Naturalist. Glasgow

Glasnik Higijenskog instituta.
Beograd

Glasnik Muzeja srpske zemlje.
Beograd.

  Ser. B. Bioloske nauk

Gloxinian. Gray, Okla.

Godisen zbornik na. Zemjodelsko-
sumarski fakultet na Univerzitet
Skopje

  Sumarstvo
  Zemjodelstvo

Godishnik na Sofiiskiya universitet
Sofiya

Godisnik. Zumarski institut.
Skopje

Godisnik na vissija finansovo-
stopanski institut. Svistov

Godisnjak Biologskog instituta u
Sarajevu. Sarajevo

Goodfruit Grower

Good Packaging. San Francisco

Gordian. Hamburg

Gorsko stopanstvo. Sofiya

Gorskostopanska Nauka. Sofiya

  Sylvicultural Science

Gorteria. Leiden

Gospodarka planova. Warszawa

Gospodarka Wodna. Warszawa

Göteborgs K. vetenskaps- och
vitterhetssamhälles handlingar.
Göteborg. Series B

Gozdarski vestnik. Ljubljana

Gradinarska i lozarska nauka.
Akademiya na selskostopanskite
nauki. Sofia

Gradinarstvo. Ministerstvo na
Selskostopanskoto Priozvodstvo.
Sofia

Graellsia. Revista de entomologos
españoles. Madrid

Grain Storage Newsletter.
FAO Rome

Grana palynologica. Stockholm

Grasas y Aceitas. Seville

Great Basin Naturalist. Provo.USA
Greenhouse, Garden, Grass.

Groenten en Fruit

Grønlandske Selskabs Aarsskrift.
Odense

Grower. London

Growth: a journal for studies of
development and increase.
Menasha, etc.

Grundförbättring. Uppsala

Grundlagen der Landtechnik.
Braunschweig

Die Grüne. Zürich

Guernsey Breeders Journal. London

Gujarat College of Veterinary
Science and Animal Husbandry
Magazine. India

Guminovye Udobreniya. Kharkov.

Guminovye Udobreniya, Teoriya i
Praktika ikh Primeneniya,
Dnepropetrovskii Sel'skokhozya-
istvennyi Institut. Kiev

Gunma Journal of Medical Sciences
Mayebaahi.

Gut. British Society of
Gastroenterology. London

Guy's Hospital Reports. London

Gynaecologia. Basel, New York

H

Half-yearly Progress Report.
Cocoa Research Institute.

Handbooks for the Identification
of British Insects. Royal
Entomological Society. London

Handbuch der Pflanzenernährung
und Dungung. Wien, New York

Hannoversche land- u. forstwirt-
schaftliche Zeitung. Hannover

Harefuah: Journal of the Medical
Association of Israel.
Jerusalem, etc.

Harper Hospital Bulletin. Detroit

Harvard Forest Papers. Petersham,
Mass.

Hassadeh. Organization of
Agricultural Labourers in
Palestine. Tel-Aviv

Hastane. Istanbul

Hausmittel über Landwirtschaft,
Europäische Wirtschaftsgemein-
schaft. Brüssel

Hautarzt. Zeitschrift fur
dermatologie, venerologie und
verwandte gebiete. Berlin, etc

Hawaii Farm Science. Honolulu

Hawaii Medical Journal and Inter-
Island Nurses Bulletin.
Honolulu

Hawaiian Planters' Record.
Honolulu

Haykakan SSR Kidovtyovnneri
Akademia, Deghekakir: Biolokiakan Kidovtyovnner

Health. Canberra, Melbourne

Health Bulletin. Edinburgh

Health Laboratory Science. USA

Health Physics. New York, London

Hedeselskabets tidsskrift. Aarhus

Hellenike kteniatrike. (Hellenic
Veterinary Medicine). Greece

Helminthologia. Academia
scientiarum slovaca. Bratislava

Helminthological Abstracts.
Commonwealth Bureau of
Helminthology. St. Albans

Helvetia Politica. Bern

Helvetica chimica acta. Basel

Helvetica medica acta. Basel

Helvetica paediatrica acta. Basel

Helvetica physiologica et
pharmacologica acta. Basel

Hemera zoa. Buitenzorg, Bogor

Herba Hungarica. Budapest

Herba polonica

Herbage Abstracts. Hurley

Hereditas, genetiskt arkiv. Lund

Heredity. An international journal of genetics. London

Herpetologica. Chicago

Hidrológiai közlöny. Budapest

Highlights of Agricultural Research Auburn

Higijena. Beograd

Hikobia. Japan

Hilgardia. A journal of agricultural science. Berkeley, Cal.

Hindustan Antibiotics Bulletin. Pimpri

Hippokrates. Stuttgart

Hirosaki Medical Journal. Hirosaki

Hiroshima Journal of Medical Sciences. Hiroshima

Hispalis medica. Sevilla

Histochemical Journal. London

Histochemie. Berlin

Hitotsubashi Journal of Economics. Tokyo

Hoard's Dairyman. Fort Atkinson, Wis.

Hodowla roślin, aklimatyzacja i nasiennictwo. Warszawa

Hofchen-Briefe. Bayer Pflanzenschutz-Nachrichten. Leverkusen

Hoja tisiológica. Montevideo

Holmbergia. Buenos Aires

Holz als Roh- und Werkstoff. Berlin

Holzforschung. Berlin

Holzforschung und Holzverwertung. Wien

Holzindustrie. Leipzig

Holzkurier. Wien

Holz. Mening

Holztechnik. Wiesbaden

Holztechnologie. Leipzig

Holzzentralblatt. Stuttgart

Holzzucht. Reinbek

Home and Garden Bulletin. US Department of Agriculture Washington

Hommes et techniques. Paris

Hommes, Terre, Eau, Rabat

Hopfen Rundschau. Wolnzach

Hoppe-Seyler's Zeitschrift für physiologische Chemie. Strassburg

Horticultura. Odense

Horticultural Abstracts. East Malling

Horticultural Advance. Horticultural Research Institute, Saharanpur. Saharanpur

Horticultural Circular. Department of Agriculture, British Columbia. Victoria, BC

Horticultural News. New Brunswick

Horticultural Research. Edinburgh

HortScience. St. Joseph, Mich.

Hospital. Revista medica de Santander. Bucaramanga

Hospital. Rio de Janeiro

Hospital de Vina del Mar. Chile

Hrana i Ishrana. Belgrade

Human Biology. Baltimore

Human Organization. Society for Applied Anthropology. New York

Human Relations. London

Humangenetik. Germany

Hungarian Agricultural Review. Budapest

Hunting Group Review. London

Hyacinth Control Journal. Fort Myers. Fla.

Hydrobiologia. Acta hydrobiologica limnologica et protistologica. Den Haag

Hygiene auf dem Lande. Berlin

Hygiène et la médecine scolaires. Paris

I.A.E.A. Occasional Papers Series.
The Institute of Asian Economic
Affairs. Tokyo

I.C.A.R. Cereal Crop Series

I.C.I. Magazine. Imperial Chemical
Industries. London

I.F.O.-Schnelldienst. Berlin

I.I.R.B. Journal of the International Institute for Sugar
Beet Research. Tienen

I.N.S.D.O.C. List: Current
Scientific Literature. Indian
National Scientific Document-
ation Centre. Delhi

I.R.I. Research Institute
New York

I.R.R.I. Reporter. International
Rice Research Institute.
Los Banos, Manila

I.T.C. Information. Delft

Ibis. A quarterly journal of
ornithology. London

Ice Cream and Frozen Confectionary
London

Ice Cream Review. Milwaukee

Idaho Agricultural Science.
Moscow, Idaho

Idia. Buenos Aires

Idia: Suplemento Forestal,
Buenos Aires

Időjárás. Budapest

Igiena, microbiologie şi
epidemiologie. Bucureşti

Igiene moderna. Genova, etc.

Igiene e sanita pubblica. Salerno

Iheringia. Series cientifica do
Museu rio-grandense de ciencias
naturais. Porto Alegre

Ikonomiceska misal. Sofija

Ikonomika selskoto stopanstvo.
Sofija

Illinois Agricultural Economics.
Urbana

Illinois Biological Monographs.
University of Illinois. Urbana

Illinois Medical Journal.
Springfield, etc.

Illinois Research. Urbana

Illinois State Florists'
Association Bulletin

Illinois Veterinarian. Urbana

Immunology. Oxford

Impact of Science on Society.
Paris

Imprensa médica. Rio de Janeiro

Impeustos. Buenos Aires

Inbred Strains of Mice. USA

Incontri.

Index to Fungi. Commonwealth
Mycological Institute. Kew

Index to Theses accepted for
Higher Degrees in the Univer-
sities of Great Britain and
Ireland. London

Index medicus. Washington

Index veterinarius. Weybridge

Indian Agriculturist. Calcutta

Indian Coconut Journal. Kerala

Indian Coffee. Bangalore

Indian Co-operative Review.
New Delhi

Indian Cotton Growing Review.
Bombay

Indian Dairyman. Bangalore

Indian Economic Journal. Bombay

Indian Economic Review.
New Delhi

Indian Farming. Delhi, etc.

Indian Food Packer. Bangalore

Indian Forest Bulletin.
New Delhi. Dehra Dun

Indian Forest Leaflet. Dehra Dun

Indian Forest Records. Dehra Dun
Entomology

Indian Forester. Dehra Dun
Indian Horticulture

Indian Journal of Agricultural
Economics. Bombay

Indian Journal of Agricultural
Science. New Delhi

Indian Journal of Agronomy.
New Delhi

Indian Journal of Animal Health

Indian Journal of Animal Science.
New Delhi

Indian Journal of Applied Chemistry
Calcutta

Indian Journal of Biochemistry.
New Delhi

Indian Journal of Chemistry
Delhi

Indian Journal of Child Health.
Bombay

Indian Journal of Dairy Science.
Bangalore

Indian Journal of Dermatology.
Calcutta

Indian Journal of Economics.
Allahabad

Indian Journal of Entomology.
New Delhi

Indian Journal of Experimental
Biology. New Delhi

Indian Journal of Genetics and
Plant Breeding. New Delhi

Indian Journal of Helminthology.
Luchnow

Indian Journal of Horticulture.
Sabour

Indian Journal of Malariology.
Calcutta

Indian Journal of Medical Research
Calcutta

Indian Journal of Medical Sciences
Bombay

Indian Journal of Meteorology and
Geophysics. New Delhi

Indian Journal of Microbiology.
Calcutta

Indian Journal of Mycological
Research. Calcutta

Indian Journal of Pathology and
Bacteriology

Indian Journal of Pediatrics.
Calcutta

Indian Journal of Pharmacy.
Benares

Indian Journal of Plant Physiology.
New Delhi

Indian Journal of Poultry Science

Indian Journal of Public Health.
Calcutta

Indian Journal of Radiology.
Madras

Indian Journal of Science and
Industry (A)

Indian Journal of Sugarcane
Research and Development.
New Delhi

Indian Journal of Surgery.
Bombay, Madras

Indian Journal of Technology.
New Delhi

Indian Journal of Veterinary
Science and Animal Husbandry.
New Delhi

Indian Labour Journal. New Delhi

Indian Medical Gazette. Calcutta

Indian Medical Journal. Bombay

Indian Pediatrics. Calcutta

Indian Physician. Bombay

Indian Phytopathology. New Delhi

Indian Practitioner. Bombay

Indian Pulp and Paper. Calcutta

Indian Science Abstracts.
Calcutta

Indian Sugar. Cawnpour, Calcutta

Indian Sugar Cane Journal.
New Delhi

Indian Tobacco. Madras

Indian Veterinary Journal.
Madras

Indiana State Board of Health
Bulletin. Indianapolis.

Indonesian Abstracts. Abstracts
on current scientific
Indonesian literature. Djakarta

Industria alimentaria. Bucureşti

Industria del Latte. Lodi

Industria lechera. Buenos Aires

Industria lemnului. Bucureşti

Industria saccarifera italiana. Genova

Industrial and Engineering Chemistry. Washington. International Edition.

Industrial and Engineering Chemistry. Fundamentals. Easton, Pa.

Industrial and Engineering Chemistry. Product Research and Development. Easton, Pa.

Industrial Medicine and Surgery. Chicago

Industrial Study. Timber Research and Development Association. London

Industrias Lacteas. Houston

Industrie Agrarie. Verona

Industrie Laitère. Paris

Industries alimentaires et agricoles. Paris

Industries de l'alimentation animale. France

Industry of Free China. Taipei

Infektsionnye zabolevaniya kul'turnykh rastenii Moldavii. Akademiva Nauk Moldavskoi SSR. Institut Fiziologii i Biokhimii Rastenii Kishinev

Informasjon fra Transportutvalget innen Skogbrukets og Skog-industrienes Forskningsforening (SSFF). Vollebekk

Információ-elektronika. Budapest

Information. Farmer Cooperative Service. US Department of Agriculture, Washington

Information. List of scientific articles published in journals. London

Information Bulletin. Near East Wheat and Barley Improved Product Project. Rome

Information Bulletin. Timber Research and Development Association. London

Information Bulletin. US Department of Agriculture. Agricultural Research Service. Washington

Information Circular. Forest Products Research, Oregon Forest Research Laboratory

Information Circular. Toxicity of Pesticides to Man. Geneva

Information Document. South Pacific Commission, Noumea. New Caledonia

Information on Land Reform, Land Settlement and Co-operatives. Rome

Information Leaflets. Inter-african Bureau for Animal Health

Information. Nitrate Corporation of Chile Limited. London

Information Report, Forest Fire Research Institute, Ottawa, Ont.

Information Report, Forest Management Research and Services Institute, Ottawa, Ont.

Information Report, Forest Products Laboratory, Vancouver

Information Report, Forest Research Laboratory, Calgary, Alberta

Information Series. Agricultural Economics, University of California, Berkeley, Cal.

Information Series. Department of Scientific and Industrial Research. Wellington, NZ.

Information Series. New Zealand Forest Service. Wellington

Informations briefe für Raumordnung, Bundesministerium des Innern, Bonn

Informations techniques. CETIOM

Informationen zur Landwirtschaft und Nahrungsgüterwirtschaft, Schwerin

Informationen zur Ökonomik Agrarwissenschaft. Berlin

Informatore fitopatologico. Bologna

Informatore di ortofrutticoltura. Bologna

Informatore zootecnico. Bologna

Informatsionnyi Byulleten' Litovskogo Nauchno-issledovatel' skogo Instituta Zemledeliya

Informatsionnyi Byulleten' Mikroelementy Sibirii

Informe. Estacion experimental de Aula Dei. Spain

Informe del Instituto Nacional de Investigaciones Agropecuarias. Ecuador

Informe mensual. Estacion experimental agricola de 'La Molina'. Lima

Informe tecnico. Instituto Forestal. Santiago

Informes Cientificos y Tecnicos. Universidad Nacional de Cuyo. Mendoza

Ingenieria agronómica. Buenos Aires

Ingenieria agronómica. Caracas

Ingenieria hidraulica en México. Mexico

Innere Kolonisation. Bonn

Insect Pest Survey. Department of Agriculture N.S.W. Sydney

Insecta matsumurana. Entomological Museum, Hokkaido Imperial University. Sapporo

Insectes sociaux. Union internationale pour l'etude des insectes sociaux. Paris

Institut Berichte, Institut für Agrarökonomik. Neetzow

Institut Francais d'oceanie. Noumea, Nouvelle Caledonie

Institut de Recherche Scientifique de Madagascar. Section de Pedologie. Tananarive-Tsimbazaza

Institute of Agricultural Research. Imperial Ethiopian Government

Institute for Agricultural Research. Samaru

Institute for Horticultural Engineering. Wageningen

Institute of Meat Bulletin. London

Institute Paper. Commonwealth Forestry Institute. Oxford

Instituto de Formento Algodonero. Bogota

Instrument and Control Engineering

Instrument Practice: Technology: Instrumentation. London

Instytut Zootechniki Wydawnictwa Wasne. Cracow

Inter-American Journal of Economic Affairs, Washington

Intercepted Plant Pests. Plant Protection Division, Production and Marketing Branch, Canada Department of Agriculture

Intereconomics. Hamburg

Interim Report. Tobacco Research Board of Southern Rhodesia. Salisbury

International Abstracts of Biological Sciences. London

International Archives of Allergy and Applied Immunology. New York, Basel

International Archives of Photogrammetry. Delft

International Atomic Energy Agency Technical Reports Series. Vienna

International Biodeterioration Bulletin. University of Aston, Birmingham

International Cat Fancy Magazine USA

International Chocolate Review. Zurich
International Clay Conference. London

International Congress for Microbiology. Montreal

International Cooperative Training Journal, Madison, Wis.

International Dairy Federation Annual Bulletin. Brussels

International Dairy Federation
International Standard. Brussels

International Development Review.
Washington

International Digest of Health
Legislation. Geneva

International Fruit World.
Basle

International Institute for Land
Reclamation and Improvement.
Wageningen

International Journal of Abstracts:
Statistical Theory and Method.
Edinburgh.

International Journal of Applied
Radiation and Isotopes.
New York, London

International Journal of
Bioclimatology and Biometeorology
Leiden

International Journal of
Biometeorology.

International Journal of Cancer.
Copenhagen

International Journal of
Immunochemistry. UK

International Journal of
Psychology. Paris

International Journal of Protein
Research. Copenhagen

International Journal of
Radiation Biology and related
Studies in Physics. London

International Journal of
Systematic Bacteriology. Ames

International Labour Review.
Geneva

International Pest Control. London

International Review of
Community Development, Rome

International Review of Cytology.
New York

International Review of Forestry
Research. New York

International Review of Tropical
Medicine

International Rice Commission
Newsletter. Bangkok

International Rice Research
Institute Reporter. Manila

International Society of Sugar
Cane Technologists, Mauritius
Congress. Amsterdam

International Sugar Journal.
Manchester, etc.

International Symposium on Forest
Hydrology. (Proceedings of a
National Science Foundation
Advanced Science Seminar, held
at the Pennsylvania State
University). New York

International Symposium on
Humidity and Moisture (1963).
Washington

Internationale Revue der gesamten
Hydrobiologie u Hydrographie.
Leipzig

Internationale Ring für
Landwirtschaft. Brüssel

Internationale Spectator.
Den Haag

Internationale Spectator.
Koninkliijk Instituut voor
Internationale Betrekkingen,
Brussel

Internationale Zeitschrift für
angewandte Physiologie,
einschliesslich Arbeitsphysiol-
ogie. Berlin

Internationale Zeitschrift der
Landwirtschaft

Internationale Zeitschrift für
Vitaminforschung. Berne and
Stuttgart

Internationaler Holzmarkt. Wien

Internationales Mykorrhizasympos-
ium. Weimar

Investigacion y Progreso
Agricola. Chile

Investigations. Department of
Agriculture, Jamaica. Kingston

Investigations. North of Scotland
College of Agriculture.
Aberdeen

Inzhenerno-Geologicheskie Svoistva
Gornykh Porod i Metody ikh
Izucheniya, Akademiya Nauk SSSR.
Otdelenie Geologo-Geografiches-
kikh Nauk. Moskva

Ion. Bucaramanga, Colombia

Iowa Farm Science. Ames

Iowa State College Journal of
Science. Ames

Iowa State Journal of Science.
Ames

Iowa State University
Veterinarian

Iowa Veterinarian. Des Moines, etc

Iranian Journal of Plant
Pathology. Tehran

Iraqi Journal of Agricultural
Science. Baghdad

Iregszemcse Bulletin

Irish Agricultural and Creamery
Review. Dublin

Irish Forestry. Wexford

Irish Journal of Agricultural
Economics and Rural Sociology.
Dublin

Irish Journal of Agricultural
Research. Dublin

Irish Journal of Medical Science
Dublin

Irish Naturalists' Journal.
Belfast.

Irish Veterinary Journal. Dublin

Irrigation and Power. Simla

Iryo. Medical Journal of the
National Hospital and Sanatorium
Tokyo. Tokyo

Israel Journal of Agricultural
Research. Rehovot

Israel Journal of Botany.
Jerusalem

Israel Journal of Chemistry.
Jerusalem

Israel Journal of Earth Sciences.
Jerusalem

Israel Journal of Experimental
Medicine. Jerusalem

Israel Journal of Medical Sciences
Jerusalem

Israel Journal of Technology.
Jerusalem

Israel Journal of Zoology

Israel Medical Journal. Haifa

Isotopes Radiation Soil-Plant
Nutrition Studies. Proceedings
of a Symposium, Ankara

Issledovaniya Fiziologii i
Biokhimii Rastenii. Institut
Eksperimental'noi Botaniki
Akademii Nauk Belorusskoi SSR

Issues and Studies. Taipei

Istanbul Üniversitesi Orman
fakültesi dergisi. Istanbul

Istanbul üniversitesi tib
fakültesi mecmuasi. Istanbul

Istanbul Üniversitesi Yayinlari
(Orman Fakültesi). Istanbul

Istorija SSSR. Moskva

Italia agricola. Piacenza

Italia forestale e montana.
Firenze

Italian Journal of Biochemistry.
Rome

Izmenenie Pochvy pri Okyl'turivanii
ikh Klassifikatsiya i Diagnostika 'Kolos'.

Izvestiya Akademii nauk
Azerbaidzhanskoi SSR. Baku

Biologicheskikh i Meditsinkikh
Nauk
Biologicheskikh i Sel'skok-
hozyaistvennikh Nauk
Geologo-Geograficheskikh Nauk

Izvestiya Akademii nauk
Belorusskoi SSR. Minsk

Biologicheskikh Nauk
Sel'sko-Khozyaistvennye Nauk

Izvestiya Akademii nauk Estonskoi
SSR. Tallinn

Biologii
Biologicheskaya

Izvestiya Akademii nauk Kazakhskoi
SSR. Alma-Ata

Biologiya
Biologicheskikh Nauk
Botaniki i Pochvovedeniya
Fiziologii i Meditsini

Izvestiya Akademii nauk Kirgizskoi
SSR. Frunze

Biologicheskikh Nauk

Izvestiya Akademii nauk
Latviiskoi SSR. Riga

Izvestiya Akademii nauk
Moldavskoi SSR. Kishinev

Biologischeskikh i Khimicheskikh
Nauk
Zoologicheskaya

Izvestiya Akademii nauk SSSR.
Leningrad

Biologiceskaja
Biologiya
Geograficheskaya i Geofiziche-
skaya
Khimicheskaya

Izvestiya Akademii nauk
Tadzhikskoi SSR. Stalinabad

Otdelenie Biologicheskikh Nauk
Otdelenie Estestvennykh Nauk

Izvestiya Akademii nauk
Turkmenskoi SSR. Ashkhabad

Biologicheskikh Nauk

Izvestiya Akademii nauk
Uzbekskoi SSR. Tashkent

Biologicheskaya

Izvestiya na Botanicheskiya
institut. Sofiya

Izvestiya Estestvenno-Nauchnogo
Instituta Permskogo Universiteta. Molotov

Izvestiya na fizicheskaya
institut s aneb, Bulgarska
Akademiya na naukite. Sofiya

Izvestiya gosudarstvennogo
nauchnoissledovatel'skogo
instituta Ozernogo i rechnogo
rybnogo khozyaistva

Izvestiya na ikonomiceskiya
Institut. Sofija

Izvestiya no Instituta po
Fiziologiya na Rasteniyata
"Metodii Popov". Bulgarska
Akademiya no Naukite. Sofia

Izvestiya na instituta po
gidrotekhnika i melioratsii,
akademiya no selskostopanskite
nauki. Sofiya

Izvestiya na instituta za gorata.
Sofiya.

Izvestiya na instituta po Obshta
i neorganichna khimiya,
Bulgarski akademiya na naukite.
Sofiya

Izvestiya na instituta po
Pshenitsata i slunchogleda
krai Gr. Tolbukhin. Sofiya

Izvestiya na instituta po
rastenievudstvo. Akademiya
selskostopanskite nauki. Sofiya

Izvestiya na instituta po
sravitelna patologiya na
zhovotnite. Bulgaria

Izvestiya na instituta po
tyutyuna. Plovdiv

Izvestiya na instituta po
zelenchukovi kulturi "Maritsa".
Plovdiv

Izvestiya na Instituta za
shivotnovudstvo "Georgi
Dimitrov". Kostinbrod. Sofiya

Izvestiya irkutskogo sel'skokhozyaistvennogo instituta.
Irkutsk

Izvestiya Kasanskogo filiala
akademii nauk SSSR. Kazan

Biologicheskikh nauk

Izvestiya Kirgizskogo filiala.
Vsesoyuznogo abshchestva
pochvovedov. Frunze

Izvestiya Kirgizskogo nauchno-
issledovatel'skogo instituta
zemledeliya. Frunze

Izvestiya Komi Filiala. Vsesoyuznogo Geograficheskogo
Obshchestva

Izvestiya Kuibyshevskogo Sel'kok-
hozyaistvennogo Instituta.
Kuibyshev

Izvestiya na Meditsinskite
instituti. Sofiya

Izvestiya na Mikrobiologicheskiya
institut. Sofiya

Izvestiya Ministerstva proizvod-
eniya i zagotovleniya sel'skok-
hozyaistvennyk produktov
Armyanskoi SSR

Izvestiya Moldavskogo filiala,
Akademiya nauk SSSR. Kishinev

Izvestiya, Nauchnoizsledovatel'ski
Institut po Mlechna Promishlenost. Vidin

Izvestiya sel'skokhozyaistvennye
Nauki. Ministerstvo Sel'skogo
Khozyaistva Armyanskoi SSR.
Erevan

Izvestiya Selskostopanski
Nauchnoiszledovatelski
institut Karnobat

Izvestiya Sibirskogo otdeleniya
Akademii Nauk. SSSR. Novosibirsk

Biologo-Meditsinskikh Nauk

Izvestiya Tikhookeanskogo
nauchnoissledovatel'skogo
instituta rȳbnogo khozyaistva i
okeanografii. Vladivostok

Izvestiya Timiryazevskoi sel'skokhozyaistvennoi akademii.
Moskva

Izvestiya na Tsentralnata
Khelmintologichna laboratoriya.
Sofiya

Izvestiya na Tsentralniya Nauchno-
Iszledovatelski Institut po
Pochvoznanie i agrotekhnika
"Nikolya Pushkarov". Sofia

Izvestiya na Tsentralniya
Veterinaren Institut na Zarazni
i Parazitni Bolesti. Bulgaria

Izvestiya na Veterinarniya
Institut za Zarazni i Parasitni
Bolesti. Sofia

Izvestiya Vsesoyuznogo geograficheskogo obshchestva. Leningrad

Izvestiya Vysshikh uchebnykh
zavedenii. Moskva

Khimiya i Khimicheskaya
Tekhnologiya
Pishchevaya Tekhnologiya
Lesnoi Zhurnal

Izvestiya na Zoologicheskiya
institut (s Muzei). Sofiya

J

J.A.R.Q. Japan Agricultural
Research Quarterly. Tokyo

J.N.K.V.V. Research Journal. India

J.P.L. Jet Propulsion Laboratory
Space Programs Summary.
(California Institute of
Technology). Pasadena

J.P.L. Jet Propulsion Laboratory.
Technical Report. (California
Institute of Technology).
Pasadena

Jaarboek. Instituut voor bewaring
en verwerking van landbouw-
produkten

Jaarboek. Instituut voor biologisch
en scheikundig onderzoek van
landbouwgewassen. Wageningen

Jaarboek. Proefstation voor de
boomwekerij te Boskoop. Boskoop

Jaarboek van de Stichting
Nederlands graancentrum.
Wageningen

Jaarboekje van het Nationaal
Instituut voor Brouwgerst, Mout
en Bier

Jaarverslag, Fakulteit van Bosbou,
Universiteit van Stellenbosch.
South Africa

Jaarverslag. Instituut voor
cultuurtechniek en waterhuishoud-
ing. Wageningen

Jaarverslag. Instituut voor
plantenziektenkundig onderzoek.
Wageningen

Jaarverslag. Instituut voor
toegepast biologisch onderzoek
in de natuur

Jaarverslag. Instituut voor
tuinbouwtechniek. Wageningen

Jaarverslag. Instituut voor
veevoedingsonderzoek 'Hoorn'.
Netherlands

Jaarverslag. Landbouwproefstation.
Suriname

Jaarverslag Landsbosbeheer.
Paramaribo

Jaarverslag. Proefstation voor de
Bloemisterij in Nederland te
Aalsmeer

Jaarverslag. Proefstation voor de
fruitteelt in de volle grond.
Wilhelminadorp

Jaarverslag. Proefstation voor de
groente- en fruitteelt onder
glas te Naaldwijk

Jaarverslag. Proefstation voor de
groenteteelt in de voole grond
in Nederland. Alkmaar

Jaarverslag. Proeftuin voor de
bloementeelt to Aalsmeer

Jaarverslag Staatsbosbeheer.
Utrecht

Jaarverslag. Stichting Bosbouwproefstation "De Dorschkamp".
Wageningen

Jaarverslag. Stichting Nederlands
Uien-Federatie. Middelharnis

Jaarverslag. Stichting voor de
Nederlandse Vlasteelt en
Vlasbewerking te Wageningen

Jaarverslag. Tuinbouwkundig
onderzoek. 's Gravenhage

Jahrbuch des Oberösterreichischen
Musealvereines. Linz

Jahrbuch für Sozialwissenschaft.
Göttingen

Jahrbuch für Tierernährung und
Fütterung. Berlin

Jahresbericht. Bayerische
Landesanstalt fur Bodenkultur,
Pflanzenbau und Pflanzenschutz.
Munchen

Jahresbericht der Biologischen
Bundesanstalt für Land- und
Forstwirtschaft in Braunschweig.
Braunschweig

Jahresbericht des Deutschen
Pflanzenschutzdienster. Hrgs.
von der Biologischen Bundesanstalt fur Land- und
Forstwirtschaft. Braunschweig

Jahresbericht der Pflanzenschutz-
zamter. Biologische Bundesanstalt für Land- und Forstwirtschaft in Braunschweig.
Stuttgart

Jahresbericht. Staatliche Lehr-
und Forschungsanstalt für
Gartenbau, Weihenstephan

Jahresbericht. Staatsinstitut für
angewandte Botanik. Hamburg

Japan Quarterly, Tokyo

Japan Science Review. Tokyo
Biological Sciences

Japanese Journal of Animal
Reproduction. Tokyo

Japanese Journal of Applied
Entomology and Zoology.
Nishigahara, Tokyo

Japanese Journal of Bacteriology.
Tokyo

Japanese Journal of Botany.
Tokyo

Japanese Journal of Breeding.
Tokyo

Japanese Journal of Dairy Science
Sendai

Japanese Journal of Ecology.
Sendai

Japanese Journal of Experimental
Medicine. Tokyo

Japanese Journal of Genetics.
Tokyo

Japanese Journal of Ichthyology.
Tokyo

Japanese Journal of Medical
Science and Biology. Tokyo

Japanese Journal of Medicine.
Tokyo

Japanese Journal of Microbiology.
Tokyo

Japanese Journal of the Nation's
Health. Kyoto

Japanese Journal of Nutrition.
Tokyo

Japanese Journal of Parasitology
Tokyo

Japanese Journal of Pharmacology
Kyoto

Japanese Journal of Physiology
Nagoya

Japanese Journal of Public Health

Japanese Journal of Sanitary
Zoology. Tokyo

Japanese Journal of Tropical
Agriculture

Japanese Journal of Veterinary
Research. Sapporo

Japanese Journal of Veterinary
Science. Tokyo

Japanese Journal of Zoology.
Tokyo

Japanese Journal of Zootechnical
Science. Tokyo

Japanese Poultry Science

Japanese Safety Forces Medical
Journal. Tokyo

Japanese Sociological Review
Tokyo

Jardins de France. Paris

Jardins et Logis

Jardins du Maroc. Casablanca

Jensal Journal. Kansas City

Jersey Cow. London

Jogtudományi Közlöny. Budapest

John Hopkins Medical Journal
Baltimore. Md.

Joint Project. Commission for
Technical Co-operation in
Africa. London

Jord og Avling. Norway

Jordbruksekonomiska Meddelanden.
Stockholm

Jord-gröda-djur. Sweden

Jornada medica. Buenos Aires

Jornal do médico. Lisboa

Jornal de pediatria. Rio de Janeiro

Jornal da Sociedade das sciencias
médicas de Lisboa

Journal of Administration
Overseas. London

Journal. Agricultural Association
of China.

Journal of the Agricultural
Chemical Society of Japan. Tokyo

Journal of the Agricultural
College. Research Institute
Kanke

Journal of Agricultural Economics.
Manchester

Journal of Agricultural
Engineering Research. Silsoe

Journal of Agricultural and Food
Chemistry. Easton, Washington

Journal of Agricultural
Meteorology. Tokyo

Journal of Agricultural Research,
Ranchi Agricultural College.
India

Journal of Agricultural Science.
Cambridge

Journal of Agricultural Science.
Tokyo

Journal of the Agricultural
Society of Finland

Journal of the Agricultural
Society of Trinidad and Tobago.
Port of Spain

Journal of the Agricultural
Society, University College of
Wales. Aberystwyth

Journal of Agriculture and Forestry
Taichung

Journal of Agriculture of South
Australia. Adelaide

Journal d'agriculture tropicale
et de botanique appliquée.
Paris

Journal of Agriculture of the
University of Puerto Rico.
Rio Pedras

Journal of Agriculture. Victorian
Department of Agriculture.
Melbourne

Journal of Agriculture of Western
Australia. Perth

Journal of the Air Pollution
Control Association. Wilmerding

Journal of the Alabama Academy
of Science. Birmingham,
Montevallo

Journal of the Albert Einstein
Medical Center. Philadelphia

Journal of Allergy. St. Louis

Journal of the American Chemical
Society. Easton

Journal of the American Dietetic
Association. Baltimore, etc.

Journal of the American Geriatrics
Society. Baltimore, Chicago

Journal of the American Medical
Association. Chicago

Journal of the American Oil
Chemists Society. Chicago, etc.

Journal of the American
Pharmaceutical Association.
Scientific Edition

Journal of the American Society of
Sugar Beet Technologists. Fort
Collins, Salt Lake City

Journal of the American Statistical
Association. Boston, etc.

Journal of the American Veterinary
Medical Association. Ithaca, NY

Journal of the American Veterinary
Radiology Society, USA

Journal of the American Water
Works Association. Baltimore.

Journal of Anatomy. London

Journal of Animal Ecology.
Oxford, Cambridge

Journal of Animal Morphology and
Physiology. Bombay, Baroda

Journal of Animal Production of
the United Arab Republic. Egypt

Journal of Animal Science.
Menasha, etc.

Journal of Antibiotics. Tokyo

Ser. A. English

Journal of Apicultural Research
London

Journal of Applied Bacteriology.
Reading

Journal of Applied Chemistry.
London

Journal of Applied Ecology.
Oxford

Journal of Applied Meteorology.
Lancaster Pa.

Journal of Applied Physics.
Lancaster Pa.

Journal of Applied Physiology.
Washington

Journal of Applied Probability. UK

Journal of the Arab Veterinary
Medical Association. UAR

Journal of the Arizona Academy of
Science. Tucson

Journal of the Arnold Arboretum.
Harvard University. Lancaster Pa.

Journal for Asian Studies.
Ann Arbor, Michigan

Journal of the Association of
Official Analytical Chemists.
Washington

Journal of the Association of
Physicians of India. Lucknow.

Journal of the Association of
Public Analysts. London

Journal of Atherosclerosis
Research. Amsterdam

Journal of the Australian Institute
of Agricultural Science.
Sydney, Melbourne

Journal of Bacteriology.
Baltimore

Journal belge de radiologie.
Bruxelles

Journal belge d'urologie.
Bruxelles

Journal of the Bengal Natural
History Society. Darjeeling

Journal of Biochemistry. Tokyo

Journal of Biological Chemistry.
Baltimore

Journal of the Biological
Photographic Association.
Baltimore

Journal of Biological Sciences.
Bombay

Journal of Biology. Osaka City

Journal. Blackface Sheep Breeders'
Association. UK

Journal of the Bombay Natural
History Society. Bombay

Journal of Bone and Joint Surgery.
Boston

A. American Volume
B. British Volume

Journal of Botany of the United
Arab Republic. Cairo

Journal of the British Grassland
Society. Hurley

Journal of the California
Horticultural Society.
San Francisco

Journal of Cell Biology. New York

Journal of Cell Science

Journal of Cellular Physiology

Journal of the Central
Agricultural Experiment Station.
Konosu

Journal of the Chartered Land
Agents' Society. London

Journal of Chemical Physics.
Lancaster, Pa.

Journal of the Chemical Society.
London

Journal of the Chemical Society
of Japan. Tokyo

Journal of Chemistry of the
United Arab Republic. Cairo

Journal of the Chiba Medical
Society

Journal of the Chinese Agricultural
Chemical Society

Journal de chirurgie. Paris

Journal of the Christian Medical
Association of India, Burma
and Ceylon. Mysore

Journal of Chromatography.
Amsterdam, London

Journal of Chromatographic
Science. Evanston

Journal of Chronic Diseases.
Oxford

Journal of Clinical Endocrinology
and Metabolism. Springfield,Ill.

Journal of Clinical Investigation.
Baltimore. etc.

Journal of Clinical Pathology.
London

Journal of the College of Arts
and Sciences, Chiba University.
Natural Science Series

Journal of the College of General
Practitioners and Research
Newsletter. London, etc.

Journal of Colloid and Interface
Science. New York, etc.

Journal of the Colorado-Wyoming
Academy of Science. Boulder

Journal of Common Market Studies.
Oxford

Journal of Comparative Neurology.
Philadelphia

Journal of Comparative Pathology.
Edinburgh, London, Croydon

Journal of Comparative and
Physiological Psychology.
Baltimore

Journal du Conseil. Conseil
permanent international pour
l'exploration de la mer.
Copenhague

Journal of cooperative extension.
Menasha, Wis.

Journal of the Czechoslovakian
Geographical Society.

Journal of Dairy Research.
London, etc.

Journal of Dairy Science.
Baltimore, Lancaster, Pa.

Journal of Dental Research.
Baltimore

Journal of the Department of
Agriculture, New Zealand.
Wellington

Journal of the Department of
Agriculture and Fisheries,
Republic of Ireland. Dublin

Journal of the Department of
Agriculture of South Australia.
Adelaide

Journal of the Department of
Agriculture for Western
Australia. Perth

Journal of Developing Areas.
Macomb, Ill.

Journal of Development Studies.
London

Journal of the Dietetic
Association, Victoria. Melbourne

Journal of Documentation. London

Journal of the Durham School of
Agriculture. Houghall

Journal of the East African
Natural History Society and
Coryndon Museum. Nairobi

Journal of Ecology. London

Journal of Economic Entomology.
Baltimore

Journal of Economic History.
New York

Journal of Economic Studies.
Oxford

Journal of the Egyptian Endocrine
Society

Journal of the Egyptian Medical
Association. Cairo

Journal of the Egyptian Public
Health Association. Cairo

Journal of the Egyptian Veterinary
Medical Association. Dar El
Hekma

Journal of the Elisha Mitchell
Scientific Society. Chapel Hill.

Journal of Embryology and
Experimental Morphology. Oxford

Journal of Endocrinology. Oxford,
Cambridge

Journal of the Entomological
Society of Australia N.S.W.
Five Dock

Journal of the Entomological
Society of British Columbia

Journal of the Entomological
Society of Queensland. Brisbane

Journal of the Entomological
Society of Southern Africa.
Pretoria

Journal of Entomology and Zoology.
Pomona College, Claremont, Cal.

Journal of Experimental Biology.
Cambridge, UK.

Journal of Experimental Botany.
Oxford

Journal of Experimental Medicine.
New York

Journal of Experimental Zoology.
Philadelphia

Journal of the Faculty of
Agriculture, Hokkaido (Imperial)
University. Sapporo

Journal of the Faculty of
Agriculture, Iwate University.
Morioka

Journal of the Faculty of
Agriculture, Kyushu University.
Fukuoka

Journal of the Faculty of
Agriculture, Shinshu University.
Ina

Journal of the Faculty of
Agriculture, Tottori University.
Tottori

Journal of the Faculty of
Fisheries and Animal Husbandry,
Hiroshima University. Fukuyama

Journal of the Faculty of
Medicine. Baghdad

Journal of the Faculty of
Radiologists. London, Bristol.

Journal of the Faculty of Science,
Hokkaido University. Sapporo

Series 5. Botany
Series 6. Zoology

Journal of the Faculty of Science,
Tokyo University. Tokyo

Series 3. Botany
Series 4. Zoology

Journal of the Farmers' Club.
London

Journal of Fermentation
Technology. Osaka

Journal of the Fisheries Research
Board of Canada. Ottawa

Journal of the Florida Medical
Association. Jacksonville

Journal of the Food Hygienic
Society of Japan. Tokyo

Journal of Food Science.
Champaign, Ill

Journal of Food Science and
Technology. Mysore

Journal of Food Science and
Technology. Tokyo

Journal of Food Technology.
Oxford

Journal of Forestry. Washington

Journal of the Forestry Commission.
London

Journal of the Formosan Medical
Association. Taipei

Journal français de médecine et
chirurgie thoracique. Paris

Journal of the Franklin Institute.
Philadelphia

Journal of Gakugei Tokushima
University. Natural Science

Journal of General and Applied
Microbiology. Tokyo

Journal of General Chemistry of
the USSR

Journal of General Microbiology.
Cambridge. UK

Journal of General Physiology.
New York

Journal of General Virology.
London

Journal of Genetic Psychology.
Provinceton

Journal of Genetics. Cambridge,
Calcutta

Journal of Geography. Chicago

Journal of the Geological
Society of Australia. Adelaide

Journal of Geology. Chicago, etc.

Journal of Geophysical Research.
Baltimore, etc.

Journal of Gerontology.
Springfield, Ill.

Journal of Helminthology. London

Journal of Heredity. Washington

Journal. Hiroshima Medical
Association. Hiroshima

Journal of Histochemistry and
Cytochemistry. Baltimore

Journal of the History of Medicine
and Allied Sciences. New York

Journal of the Horticultural
Association of Japan. Tokyo

Journal of Horticultural Science.
London

Journal of Hydrology. Amsterdam

Journal of Hygiene. Cambridge

Journal of Hygiene, Epidemiology,
Microbiology and Immunology.
Prague

Journal of Hygienic Chemistry
(Eisei Kagaku) Tokyo

Journal of Immunology. Baltimore

Journal of the Indian Botanical
Society. Madras

Journal of the Indian Chemical
Society. Calcutta

Journal of the Indian Institute
of Science. Bangalore

Journal of the Indian Medical
Association. Calcutta

Journal of the Indian Society of
Agricultural Statistics.
New Delhi

Journal of the Indian Society of
Soil Science. New Delhi

Journal of the Indiana State
Medical Association. Indianapolis

Journal of Industrial Economics.
Oxford

Journal of Infectious Diseases.
Chicago

Journal of Insect Physiology,
Oxford

Journal of the Institute of
Animal Technicians. London

Journal of the Institute of
Bankers. London

Journal of the Institute of
Brewing. London

Journal of the Institute for
Scientific and Technical
Information. Prague

Journal of the Institute of Wood Science. London

Journal of the Institution of British Agricultural Engineers. London

Journal of Institution of Chemists. India. Calcutta

Journal of Inter-American Studies. Gainesville, Fla.

Journal international de chirurgie. Bruxelles

Journal of the International College of Surgeons. Chicago, etc.

Journal of the International Institute of Sugar Beet Research.

Journal of Invertebrate Pathology. USA

Journal of Investigative Dermatology. Baltimore

Journal of the Iraqi Medical Professions. Baghdad

Journal of the Irish Grassland and Animal Production Association.

Journal of the Iwate Medical Association. Japan

Journal of the JJ Group of Hospitals and Grant Medical College. Bombay

Journal of the Jamaican Association of Sugar Technologists

Journal of the Japan Veterinary Medical Association. Tokyo

Journal of the Japan Wood Research Society. Tokyo

Journal of the Japanese Association of Rural Medicine

Journal of Japanese Botany. Tokyo

Journal of the Japanese Forestry Society. Tokyo

Journal of the Japanese Society of Food and Nutrition. Tokyo

Journal of the Japanese Society of Grassland Science.

Journal of the Japanese Society for Horticultural Science.

Journal of the Kansas Entomological Society. Manhattan.

Journal of the Kansas Medical Society. Columbus

Journal of the Kanto-Tosan Agricultural Experiment Station. Konosu, Saitama

Journal of the Kentucky State Medical Association. Louisville

Journal of the Kumamoto Medical Society

Journal of the Kurume Medical Association. Kurume

Journal of the Kyoto Prefectural Medical University

Journal of Labelled Compounds. Brussels

Journal of Laboratory and Clinical Medicine. St. Louis. Mo.

Journal of the Land Agents' Society. London

Journal of Laryngology and Otology London

Journal of the Linnean Society. London

Botany
Zoology

Journal of Lipid Research. Memphis, New York

Journal of Local Administration Overseas. London

Journal of the Louisiana State Medical Society. New Orleans

Journal of the Madras University. Madras

Section B. Contributions in Mathematics, Physical and Biological Sciences

Journal of the Maharaja Sayajirao University of Baroda. Baroda

Journal of the Maine Medical Association. Portland

Journal of the Malayan Veterinary Medical Association. Singapore

Journal of Mammalogy. Baltimore.

Journal of the Marine Biological Association of India. Mandapam Camp, Madras

Journal of the Marine Biological Association of the United Kingdom. Plymouth

Journal of Marketing. Chicago

Journal de médecine de Bordeaux (et de la region du sud-ouest)

Journal de médecine de Lyon

Journal of the Medical Association of Georgie. Augusta

Journal of the Medical Association of the State of Alabama. Montgomery

Journal of the Medical Association of Thailand

Journal of Medical Entomology. Honolulu

Journal of Medical Laboratory Technology. London

Journal médical libanais. Beirut

Journal of Medical Microbiology, Edinburgh

Journal of the Medical Society of New Jersey. Newark, etc.

Journal of Medicinal Chemistry. USA

Journal of Mental Deficiency Research. Sevenoaks

Journal of Mental Science. London

Journal of the Michigan State Metical Society. Detroit

Journal de microscopie.

Journal of Microscopy. London

Journal of Milk and Food Technology. Albany, NY.

Journal of the Ministry of Health Cairo

Journal of the Mississipi Academy of Sciences Inc. State College, Mississipi

Journal of Modern African Studies. London

Journal of Molecular Biology. London, New York

Journal of Morphology. Philadelphia, etc.

Journal of the Mount Sinai Hospital New York, etc.

Journal of Nagoya City University Medical Association.

Journal of Nagoya Medical Association

Journal of Nara Gakugei University

Journal of the Nara Medical Association. Kashihara

Journal of the National Agricultural Society of Ceylon

Journal of the National Cancer Institute. Washington, etc.

Journal. National Institute of Agricultural Botany. Cambridge

Journal of the National Medical Association. Tuskegee, Ala.

Journal. National Research Council of Thailand. Bangkok

Journal of Natural History. London

Journal of Nervous and Mental Diseases. New York

Journal of Neurochemistry. London, New York

Journal of Neurology, Neurosurgery and Psychiatry. London

Journal of Neuropathology and Clinical Neurology. Chicago

Journal of Neuropathology and Experimental Neurology. Baltimore

Journal of Neurophysiology. Springfield

Journal of Neurosurgery. Springfield

Journal of the New Zealand Dietetic Association Incorporated. Wellington

Journal of the Nigerian Institute for Oil Palm Research. Benin City

Journal of Nuclear Medicine. Chicago

Journal of Nutrition.
Baltimore, etc.

Journal of Nutrition and
Dietetics. Coimbatore

Journal of Obstetrics and
Gynaecology of the British
Commonwealth. London

Journal officiel de la republique
francaise, avis et rapports du
conseil economique et social.
Paris

Journal of the Okayama Medical
Society. Okayama

Journal of the Oklahoma State
Medical Association. Muskogee

Journal of Organic Chemistry.
Baltimore

Journal of Organic Chemistry
of the USSR

Journal of the Osaka City Medical
Center. Osaka

Journal of the Osaka Medical
College.

Journal of the Oslo City Hospitals
Oslo

Journal of Osteopathy.
Philadelphia

Journal of the Oxford University
Forest Society. Oxford

Journal of the Pakistan Medical
Association. Karachi

Journal of Parasitology. Lancaster
Pa.

Journal of Pathology. London

Journal of Pediatrics. St. Louis.

Journal of Pharmaceutical
Sciences.

Journal of the Pharmaceutical
Society of Japan. Tokyo

Journal of Pharmacology and
Experimental Therapeutics.
Baltimore

Journal of Pharmacy and
Pharmacology. London

Journal of the Philippine
Medical Association. Manila

Journal of Physical Chemistry.
Ithaca

Journal de physiologie. Paris

Journal of Physiology. London

Journal of Political Economy.
Chicago

Journal of Postgraduate Medicine.
Bombay

Journal of the Postgraduate
School, Indian Agricultural
Research Institute. New Delhi

Journal fur praktische Chemie.
Leipzig

Journal and Proceedings of the
Institute for Sewage Purification
Kew

Journal and Proceedings of the
Institution of Agricultural
Engineers. London

Journal and Proceedings of the
Institution of Chemists,
India. Calcutta

Journal and Proceedings of the
Royal Society of New South
Wales. Sydney

Journal of Protozoology.
Utica, NY

Journal of the Public Health
Association of Japan. Tokyo

Journal of the Quekett
Microscopical Club. London

Journal de radiologie,
d'électrologie et de médecine
nucléaire. Paris

Journal of Range Management.
Portland, Or.

Journal of Reproduction and
Fertility. Oxford

Journal of Research. Punjab
Agricultural University.
Ludhiana.

Journal of the Rio Grande Valley
Horticultural Society. Weslaco

Journal of the Royal Agricultural
Society of England. London

Journal of the Royal Army Medical
Corps. London, etc.

Journal of the Royal Army
Veterinary Corps. Aldershot

Journal of the Royal Association
of British Dairy Farmers.
London

Journal of the Royal Caledonian
Horticultural Society.
Edinburgh

Journal of the Royal College of
Physicians of London, UK

Journal of the Royal Horticultural
Society. London

Journal of the Royal Microscopical
Society. London

Journal of the Royal Naval
Medical Service. London

Journal of the Royal Sanitary
Institute. London

Journal of the Royal Society of
Arts. London

Journal. Royal Society of
Western Australia. Perth

Journal of the Royal Statistical
Society. London

Series A. General

Journal of the Rubber Research
Institute of Malaya. K.Lumpur

Journal of Rural Development and
Administration. Peshawar

Journal of the Sanitary Engineer-
ing Division. Proceedings of the
American Society of Civil
Engineers. New York

Journal of the Science of Food
and Agriculture. London

Journal of Science of the
Hiroshima University. Hiroshima

Series B. Division 1. Zoology

Journal of the Science of Labour
Tokyo

Journal of the Science of Soil
and Manure. Tokyo

Journal des sciences médicales de
Lille.

Journal of the Scientific
Agricultural Society of Finland.
Helsinki

Journal of Scientific and Industrial Research. New Delhi

Journal of Scientific Instruments.
London

Journal of Scientific Research
of the Banaras Hindu University.

Journal of Sedimentary Petrology.
Tulsa, Okla.

Journal of Sericultural Science.
Tokyo

Journal of the Showa Medical
Association

Journal of Small Animal Medicine.
Inglewood, Calif.

Journal of Small Animal Practice.
Oxford, etc.

Journal de la Société centrale
d'agriculture de Belgique.
Bruxelles

Journal of the Society for
British Entomology. Southampton

Journal of the Society of Dairy
Technology. London

Journal of the Society of Textile
and Cellulose Industry, Japan.
Tokyo

Journal of the Soil Conservation
Service of New South Wales.
Sydney

Journal of Soil Science. Oxford

Journal of the Soil Science
Society of the Philippines.
Manila

Journal of Soil Science of the
United Arab Republic. Cairo

Journal of Soil and Water
Conservation. Baltimore, etc.

Journal of Soil and Water
Conservation in India.
Hazaribagh

Journal of South African
Biological Science.

Journal of South African Botany.
Cape Town

Journal of the South African
Veterinary Medical Association.
Onderstepoort

Journal of the Sports Turf
Research Institute. Bingley

Journal of Stored Products
Research. Oxford

Journal of Taiwan Agricultural
Research. Taipei

Journal of the Tennessee Academy
of Science. Nashville

Journal of the Tennessee State
Medical Association. Nashville

Journal of the Thai Veterinary
Medical Association

Journal of Theoretical Biology.
UK

Journal of Thoracic and Cardiovascular Surgery. St. Louis

Journal of the Timber Development
Association of India. Dehra Dun

Journal of the Tokyo Medical
College.

Journal of the Tokyo Society of
Veterinary and Zootechnical
Science.

Journal of the Tokyo University
of Fisheries. Yokosuka City

Journal of Tropical Geography.
Singapore

Journal of Tropical Medicine and
Hygiene. London

Journal of Tropical Pediatrics.
New Orleans, London

Journal of Ultrastructure
Research. New York, London

Journal of the University of
Bombay. Fort Bombay.

　Section A. Biological Sciences
　Section B. Physical Sciences

Journal of the University of
Newcastle-upon-Tyne, Agricultural Society.

Journal of the University of
Saugar. Saugar. Part II,
Section B

Journal d'urologie medicale et
chirurgicale. Paris

Journal of Urology. Baltimore

Journal of Veterinary and Animal
Husbandry Research. Mhow

Journal of Veterinary Medicine.
Tokyo

Journal of Veterinary Science of
the United Arab Republic.
Egypt

Journal of Virology. Baltimore

Journal of Vitaminology. Kyoto

Journal of the Washington Academy
of Sciences. Washington

Journal of the Water Pollution
Control Federation. Washington

Journal of the West African
Science Association. Achimota,
London.

Journal of Wildlife Management.
Menasha

Journal of the Yokohama Municipal
University. Yokohama

Journal of the Zoological Society
of India. Calcutta

Journal of Zoology. Proceedings
of the Zoological Society of
London

Journal-Lancet. Minneapolis

Jugoslavenska Pedijatrija. Zagreb

Jugoslovenski Pregled. Beograd

Jugoslovensko drustvo za
proucavanje zemljista. Beograd

Justus Liebigs Annalen der
Chemie. Leipzig

Jute Bulletin. Calcutta

K

K.T.L. - Berichte über Landtechnik. Frankfurt a.M.

Kaffee und Teemarkt. Hamburg

Kagaku. Kyoto

Kajian Veterinaire. Malaysia

Kali. Netherlands

Kalibriefe. Hannover

Kanpur Agricultural College
 Journal. Kanpur

Kansas University Science
 Bulletin. Lawrence

Karakulevodstvo i zverovodstvo.
 Moskva

Karjantuote. Helsinki

Karjatalous. Helsinki

Karstenia. Suomen Sieni-Seura.
 Helsinki

Kartofel' i ovoshchi. Moskva

Kartoffelbau. Hamburg, etc.

Kartoffelwirtschaft. Hamburg

Kasikorn. Bangkok

Keio Journal of Medicine. Tokyo

Keizai Kenkyu. Tokyo

Kenley Abstracts. British Paper and
 Board Industry Research
 Association. Kenley

Kent Farmer. Maidstone

Kentucky Farm and Home Science.
 Lexington

Kenya Coffee. Nairobi

Kenya Dairy Farmer. Nairobi

Kenya Farmer. Nakaru

Kenya Information Services Bulletin

Kenya Sisal Board Bulletin.
 Nairobi

Kerala Vet. Mannuthy, Trichur.
 India

Kernenergie. Berlin

Kertészeti kutató intézet
 kozlemenyei

Kertészeti és Szólészet. Budapest
Kertészeti és szolészeti főiskola
 évkönyve. Budapest

Kertészeti és szolészeti főiskola
 kozlemenyei

Keszthelyi agrartudomanyi foiskola
 kiadvanyai. Budapest

Keszthelyi mezogazdasagi akademia
 kiadvanyai. Budapest

Kew Bulletin. Royal Botanic
 Gardens, Kew.

Key to Turkish Science.
 Agriculture. Ankara

Khadi gramodyog. Bombay

Kharchova Promislovist. Kiev

Khidrologiya i meteorologiya.
 Sofiya

Khimiya v sel'skom khozyaistve.
 Moskva

Khirurgiya. Moskva

Khirurgiya. Sofiya

Khlebopekarnaya i Konditerskaya
 Promyshlennost'. Moscow

Khlopkovodstvo. Moskva

Kholodil'naya tekhnika. Moskva

Khranitelna promishlenost. Sofiya
 Sofiya

Kieler Meeresforschungen. Kiel

Kieler milchwirtschaftliche
 Forschungsberichte. Hildesheim

Kieler Studien. Institut für
 Weltwirtschaft, Universität
 Kiel. Kiel

Kinderärztliche Praxis. Leipzig

Kirkia. Journal of the
 Government Herbarium.
 Salisbury, Rhodesia

Kisérletügyi közlemények.
 Budapest

Kitakanto Medical Journal

Kitasato Archives of Experimental
 Medicine. Tokyo

Kleintier-Praxis. Hannover

Kleintierzucht in Forschung und
 Lehre. Celler Jahrbuch. Celle

Kliniceskaja medicina. Moscow

Klinicheskaya meditsina. Moskva

Klinika oczna. Warszawa

Klinische medizin. Wien

Klinische Monatsblätter für
 Augenheilkunde. Stuttgart

Klinische Wochenschrift. Berlin

Klucze do oznaczania owadów
 Polski. Warszawa

Knižni a Dokumentacni Zpravodaj, Czechoslovakia

Koedoe. Journal for scientific research in the National Parks of the Union of South Africa. Pretoria

Koeltechniek. The Hague

Koleopterologische Rundschau. Wien

Kolhozno-Sovhoznoe Proizvodstvo.

Kolloidnyi zhurnal. Voronezh. Moskva

Kolloidzeitschrift. Dresden

Kommunist. Moskva

Konevodstvo (afterwards: i konnyĭi sport). Moskva

Kongelige Norske videnskabernes selskabs skrifter. Trondhjem

Kongelige Veterinaer- og Landbohøiskoles Aarsskrift. Kjøbenhavn

Konservnaya i ovoshchesushil'naya promyshlennost'. Moskva

Kontyû. Entomological Society of Japan. Tokyo

Konzerv- és Paprikaipar. Budapest

Kooperation. Berlin

Kooperativno zemedelie. Sofiya

Kora vyvetrivaniya. Institut geologii rudnykh mestorozhdenii, petrografii, i geokhimii. Moskva

Kortárs. Budapest
Korte mededelingen. Stichting bosbouwproefstation 'De Dorschkamp'. Wageningen

Koruma. Istanbul

Kosmos. Stuttgart

Kosmos. Warszawa. Seria A.Biologia

Kožařstvi. Praha

Kozgazdasagi szemle. Budapest

Krolikovodstvo (i zverovodstvo). Moskva

Kroniek van Afrika. Leiden

Kühn-Archiv. Berlin

Kukuruza. Moskva

Kultura i spoleczenstwo. Warszawa

Kulturpflanze. Berlin

Kumamoto Medical Journal. Kumamoto

Kungliga Fysiografiska sällskapets i Lund förhandlingar. Lund

Kungliga Lantbruksakademiens tidskrift. Stockholm

Kungliga Skogs- och lantbruksakademiens tidskrift. Stockholm

Kungliga Tekniska högskolans handlingar. Stockholm

Kurtziana. Argentina

Kurukshetra. New Delhi

Kurume Medical Journal. Kurume-shi

Kurz und bündig. Auslese aus den neuesten landwirtschaftlichen Veröffentlichungen. Limbergerhof

Kwartalnik historii. Warszawa

Kyklos. Bern

Kyushu Agricultural Research. Fukuoka

Kyushu Journal of Medical Sciences. Fukuoka

Kyushu Memoirs of Medical Sciences. Fukuoka

L

LABDEV. Journal of Science and Technology. Kanpur

Part B. Life Sciences

L.S.U. Forestry Notes. Louisiana State University. School of Forestry and Wildlife Management. Baton Rouge

L.S.U. Wood Utilization Notes. Louisiana State University. School of Forestry and Wildlife Management. Baton Rouge

Laboratorio. Granada

Laboratornoe delo. Moskva

Laboratory Animals. Journal of
the Laboratory Animal Science
Association. London

Laboratory Animal Care. USA

Laboratory Investigation.
International Academy of
Pathology. New York, Philadelphia

Laboratory Practice. London

Laden Wain. Agricultural Club
of the University of Reading.
Reading

Lait. Lyon, Paris

Lalahan Zootekni Arastirma
Enstitasii Dergisi. Ankara

Lal-Baugh. Journal of the Mysore
Horticultural Society.
Bangalore

Lambillionea. Revue mensuelle de
l'Union des entomologistes
belges. Bruxelles

Lancet. London

Land Economics. Madison, Wis

Land- und Forstwirtschaftlicher
Betrieb. Wien

Land and Labour. Prague

Land Reform, Land Settlement and
Co-operatives. Rome

Land Research Series. C.S.I.R.O.
Australia. Melbourne

Land Resource Study. Land
Resources Division, Directorate
of Overseas Surveys, Tolworth,
Surrey

Die Landarbeit. Stuttgart

Landbauforschung Völkenrode.
Forschungsanstalt für Landwirtschaft, Braunschweig-Völkenrode

Landbouwdocumentatie.
's Gravenhage

Landbouwgids. Utrecht

Landbouwkundig tijdschrift.
's Gravenhage, Wageningen

Landbouwmechanisatie. Amsterdam,
Wageningen

Landbouwvoorlichting.
Rijkslandbouwvoorlichtingsdienst.
's Gravenhage

Landtechnische Forschung.
Wolfratshausen

Land- en Tuinbouw Jaarboek.
Belgium

Landwirtschaft. Wien

Landwirtschaft-angewandte
Wissenschaft. Bonn

Landwirtschaftliche Forschung.
Darmstadt, etc.

Landwirtschaftliche Hochschule
Hodenheim Raden und
Abhandlungen

Landwirtschaftliche Schriftenreihe
Boden und Pflanze. Bochum

Landwirtschaftliche Zeitschrift
fur die Nord-Rheinprovinz. Bonn

Landwirtschaftliche Zeitschrift
fur die Rheinprovinz. Bonn

Landwirtschaftliches Jahrbuch der
Schweiz. Bern

Landwirtschaftliches Wochenblatt
fur Kurhessen-Waldeck

Landwirtschaftliches Wochenblatt
fur Westfalen und Lippe

Landwirtschaftliches Zentralblatt
Berlin.
Abt. 4. Veterinärmedizin

Landwirtschaftsblatt Weser-Ems.
Oldenburg

Langenbecks Archiv für Klinische
Chirurgie. Berlin

Lantbrukshögskolans annaler.
Uppsala

Lantbrukshögskolans Diss.

Lantbrukshögskolans Meddelanden.
Uppsala, Stockholm

Lantbruksveckan. Stockholm

Lantman och andelsfolk.
Helsingfors.

Lantmannen Svensktland. Stockholm

Laporan, Lembaga Penelitian
Kehutanan, Bogor, Indonesia

Las Polski. Warszawa

Lasca Leaves. Los Angeles

Latin American Business Highlights
Chase Manhattan Bank. New York

Lattante. Parma

Latte. Milan

Latvijas Lauksaimniecibas Akademijas Raksti. Latvia

Latvijas Lopkopibas un Veterinarijas Zinatniski Petnieciska Instituta Raksti. Latvia

Latvijas PSR zinātņu akademijas vestis. Riga

Laval médical. Quebec

Lavori di botanica. Istituto botanico e di fisologia vegetale dell' Universita de Padova

Lavori dell'Istituto botanico dell' Universita di Milano

Lavora arrozeira. Porto Alegre

La-yaaran. Ilanoth

Leaflet. Alabama Agricultural Experiment Station. Auburn

Leaflet. California Agricultural Experiment Station.

Leaflet. Coconut Research Institute of Ceylon. Lunawila

Leaflet. Coffee Board Research Department. Balehonnur

Leaflet. Department of Agriculture, Western Australia. Perth

Leaflet. Edinburgh and East of Scotland College of Agriculture. Edinburgh

Leaflet. Forest Products Research Laboratory. Ministry of Technology, London

Leaflet of the Forestry Commission London

Leaflet. Forestry and Timber Bureau. Canberra

Leaflet. Forestry Division, National and University Institute of Agriculture. Ilanoth

Leaflet. Ministry of Agriculture, Northern Ireland. Belfast

Leaflet. Texas Agricultural Experiment Station

Leaflet. US Department of Agriculture. Washington

Lebensmittel und Ernahrung. Wien

Lebensmittel-Industrie. Berlin

Lebensmitteltierarzt. Hannover

Lebensmittel Wissenschaft und Technologie. Zurich

Len a konopi. Vyzkumna Stanice Zemedelska v Sumperku-Temenici.

Len i konopliya. Moskva

Leprosy Review. London

Les. Bratislava

Les. Ljubljana

Lesnaya promyšhlennost'. Moskva

Lesnický časopis. Bratislava, Prague

Lesnická prace. Warszawa

Lesnoe khozyaǐstvo. Moskva

Lesnoi zhurnal. Arkhangel'sk

Lesovedenie. Moscow

Letopis nauch. Radova

Levante. Roma

Library Association Record. London

Library and Documentation Bulletin. Department of Scientific-Technical Information, University of Agriculture in Brno. Brno

Library List. National Agricultural Library, US Department of Agriculture. Washington

Library Review. Forestry Commission. London

Liebigs Annalen der Chemie. Weinheim

Lietuvos TSR mokslu akademijos Biologijos instituto darbai. Vilnius

Lietuvos TSR mokslu akademijos darbai. Vilnius

Serija C.

Lietuvos Zemdirbystes Mikslinio Tyrimo Instituto Darbai

Lietuvos zemes ukio akademijos mokslu darbai. Vilnius.

Life of Science. Krakow

Life Sciences. Oxford

Lighter. Experimental Farms
Service Tobacco Division.
Ottawa

Lille chirurgical. Lille

Lilloa. Revista de botánica.
Tucumán

Lily Yearbook

Limnologica. Germany

Limnology and Oceanography.
Baltimore

Lingnan Science Journal. Canton

Lipids. Chicago, Ill.

Listy cukrovarnické. Praha

Living Conditions and Health.
World Congress of Doctors for
the Study of Present-day
Living Conditions. Vienna, etc.

Lloydia: a quarterly journal of
biological science. Manasha, Wis

Lloyds Bank Review. London

Locusta. Kara

London and Cambridge Economic
Bulletin. Cambridge

Loodusuurijate seltsi aastaraamat.
Tallinn

Lotta antiparassitaria. Roma

Lotta contro la tubercolosi.
Roma

Loughborough Journal of Social
Studies. Loughborough

Louisiana Agriculture

Lozarstvo i Vinarstvo. Sofija

Lucrări ştiinţifice. Institutul
agronomic "Dr. Petru Groza".
Cluj. Bucuresti

Lucrări ştiinţifice. Institutul
agronomic "Professor Ion Ionescu
de la Brad", Iasi. Bucuresti

Lucrări ştiinţifice. Institutul
agronomic "Nicolae Balcescu".
Bucuresti

Lucrări ştiinţifice. Institutul
Politehnic Galaţi. Bucuresti

Lucrárile. Gradinii botanice din
Bucuresti

Lucrárile Institutului de
cercetari alimentare. Bucuresti

Lucrárile Institutului de
cercetari veterinaire si
biopreparate "Pasteur".
Roumania

Lucrárile ştiinţifice. Institutul
agronomic craiova. Bucuresti

Lucrárile ştiinţifice. Institutul
agronomic timisoara

Lucrárile ştiinţifice. Institutul
de cercetari horti-viticole.
Baneasa-Bucuresti

Lucrárile ştiinţifice ale
Institutului de cercetari
zootehnice. Bucuresti

Lucrárile ştiinţifice ale
Institutului de patologie si
igiena animala. Bucuresti

Lucrárile ştiinţifice. Institutul
Politehnic. Cluj

Luga i pastbishcha.

Lunds universitets arsskrift

Lutra. Vereniging voor zoogdier-
kunde en zoogdierbescherming.
Leiden

Lyon chirurgical. Lyon, etc.

Lyon médical. Lyon

Lyon pharmaceutique. Lyon

M

M.G.A. Bulletin. Mushroom
Growers' Association. Yaxley

M.T.A. III Osztaly Közleményei.
Budapest

Maandblad voor de Landbouwvoor-
lichtingsdienst. 's Gravenhage

Maandschrift Economie. Tilburg

Maandschrift voor kindergenees-
kunde. Amsterdam

Maandstatistiek van de Landbouw.
Zeist

Maanedsoversigt over
Plantesygdomme. Lyngby

Maanedsskrift for praktisk
Laegegerning og social Medicin.
København

Maatalous. Helsinki

Maatalous ja koetoiminta.
Helsinki

Maataloustieteelinen
aikakauskirja. Helsinki

Macchine e motori agricole.
Roma

Macdonald Farm Journal. Montreal

Maderero. Santiago, Chile

Madhya Bharati. Journal of the
University of Saugar. Saugar.

Sect. IIB Natural Science

Madjalah kedokteran Indonesia.
Djarkarta

Madras Agricultural Journal.
Madras, Coimbatore

Madras Journal of Co-operation.
Madras

Madras Veterinary College Annual

Madrono. Journal of the
California Botanical Society.
San Francisco, etc.

Maelkeritidende. Odense

Magyar allatorvosok lapja.
Budapest

Magyar mezőgazdaság. Budapest

Magyar sebészet. Budapest

Magyar tudományos akadémia
agrártudományok osztályanak
közleményei. Budapest

Maha Magazine

Maharashtra co-operative quarterly.
Bombay

Maine Farm Research. Orono

Maine Veterinarian

Maize. Udaipur

Maize Genetics Cooperation News
Letter. Ithaca, NY

Makromolekulare Chemie.
Freiburg i.B., Basle

Malacologia. USA

Maladies des plantes maraichères.
Rapport d'activité. Station de
Pathologie Végétale du Sud-Est,
Station de Pathologie Végétale
du Sud-Ouest. Montfavet,
Pont-de-la-Maye

Malaysian Agricultural Journal
Kuala Lumpur

Malaysian Agri-Horticulture
Association Magazine.
Kuala Lumpur

Malaysian Economic Review.
Singapore

Malaysian Forester. Kuala Lumpur

Malaysian Forest Records.
Kuala Lumpur

Malaysian Nature Journal.
Kuala Lumpur

Malaysian Veterinary Journal.
Malaysia

Mammalia. Morphologie, biologie,
systématique des mammifères.
Paris

Man in India. Ranchi

Management Science. Baltimore

Manchester School of Economic and
Social Studies. Manchester

Manufacturing Chemist. London

Manual. Instituto Forestal,
Santiago

Manx Journal of Agriculture.
Peel, Douglas

Marcellia. Revista internazionale
di cecidologia. Strasbourg

Marchés tropicaux et
méditerranéens. Paris

Marchés tropicaux du monde. Paris

Marine Research. Department of
Agriculture and Fisheries for
Scotland

Market Growers' Journal.
Louisville, Ky.

Marketing Report. Department of
Agricultural Economics,
University of Leeds. Leeds

Marketing Research Report.
Economic Research Service,
US Department of Agriculture,
Washington

Marketing Research Report.
Farmer Cooperative Service,
US Department of Agriculture,
Washington

Marketing Research Report.
US Department of Agriculture,
Washington

Marketing Series. Department of
Agricultural Economics,
Wye College, University of
London, Ashford

Maroc médical. Casablanca

Marseille chirurgical. Marseille

Marseille médical. Marseille

Maryland Florist

Maslozhirovaya Promyshlennost'.
Moscow

Material und Organismen. Berlin

Materialsammlung der Agrarsoziale
Gesellschaft. Göttingen

Materialȳ i Tezisȳ VI Konferentsii
po Khimizatsii Sel'skogo
Khozyaĭstva. Orenburg

Maydica. Bergamo

Mayo Clinic Proceedings.
Rochester, USA

Meat and Dairy Produce Bulletin.
London

Meat Trades Journal and Cattle
Saleman's Gazette. London

Mecanizarea si Electrificarea
Agriculturii. Bucuresti

Mechanisace zemědělství. Praha

Meddelanden. Forskningstiftelsen
skogsarbeten. Stockholm

Meddelanden från Gullåkers
växtförädlingsanstalt. Hammenhög

Meddelanden från Instituionen
för Lantbrukets byggnadstehnik,
Lantbrukshögskolan, Uppsala

Meddelanden. Jordbrukets
Utredningsinstitut. Stockholm

Meddelanden. Jordbrukstekniska
institutet. Uppsala

Meddelanden. Kungliga Lantbruk-
shogskolan (och Statens)
Lantbruksforsok (Statens)
Jordbruksforsok. Uppsala

Meddelanden från Kungl.
Veterinärhögskolan i Stockholm

Meddelanden. Lantbrukshögskolan
och Statens Lantbruksförsok.
Uppsala
Serie A
Serie B

Meddelanden från Statens centrala
frökontrollanstalt. Stockholm

Meddelanden. Statens maskinprovn-
ingar. Uppsala

Meddelanden från Statens
skogsforskningsinstitut.
Stockholm, etc.

Meddelanden från Statens
växtskyddsanstalt. Stockholm

Meddelanden från Statens
veterinärmedicinska anstalt.
Stockholm

Meddelanden. Stiftelsan för
Rasförädling av Skogsträd.
Helsinki

Meddelanden. Svenska Mejeriernas
Riksförening. Stockholm

Meddelanden. Svenska träforskning-
sinstitutet (Trakemi och
Papperstekník). Stockholm

Meddelanden. Svenska träforskning-
sinstitutet. Träteknik.
Stockholm

Meddelelser. Kongelige Veterinaer-
og Landbohøjskolen Afdeling for
Landbrugets Plantekultur. Denmark

Meddelelser om Grønland.
Kjøbenhavn

Meddelelser fra det Norske
Myrselskap. Oslo

Meddelelser fra det Norske
skogsforsøksvesen.
Oslo, Vollbekk

Meddelelser. Norsk treteknisk
institutt. Blindern

Meddelelser fra Statens
Forsøksvirksomhed i Plantekultur,
Kjøbenhavn, etc.

Meddelelser fra Vestlandets
forstlige forsøksstatjion.
Bergen

Médecine tropicale. Marseille

Mededelingen. Bedrijfsvoorlicht-
ingsdienst voor de tuinbouw in
de Provincie Oost-Vlaanderen

Mededelingen. Directeur van de
tuinbouw. 's Gravenhage

Mededelingen van de Geologische
stichting. 's Gravenhage

Mededelingen. Instituut voor
biologisch en scheikundig
onderzoek van landbouwgewassen.
Wageningen

Mededelingen. Instituut voor
cultuurtechniek en waterhuish-
ouding. Wageningen

Mededelingen. Instituut voor
plantenziektenkundig onderzoek.
Wageningen

Mededelingen van het Instituut
voor rationele suikerproductie.
Bergen-op-Zoom

Mededelingen van het Instituut
voor toegepast biologisch
onderzoek in de natuur. Arnhem

Mededelingen van het Instituut
voor de veredeling van
tuinbouwgewassen. Wageningen

Mededelingen. Laboratorium voor
entomologie. Wageningen

Mededelingen. Laboratorium voor
phytopathologie. Wageningen

Mededelingen van het Laboratorium
voor plantecologie. Rijksland-
bouwhogeschool. Gent

Mededelingen van de Landbouwhoge-
school te Wageningen

Mededelingen. Landbouwproefstation in Suriname. Paramaribo

Mededelingen van de Nederlandse
algemeene keuringsdienst voor
landbouwzaken en aardappelpoot-
goed. Wageningen

Mededelingen van de Nederlandse
algemene keuringsdienst voor
zaaizaad en pootgoed van
landbouwgewassen. Netherlands

Mededelingen. Nederlands Instituut
voor Zuivelonderzoek. Ede

Mededelingen uit het Phytopathol-
ogisch laboratorium "Willie
Commelin Scholten". Amsterdam

Mededelingen. Proefstation voor
de akker- en weidebouw.
Wageningen

Mededelingen van het Proefstation
voor de groenten- en fruitteelt
onder glas. Naaldwijk

Mededelingen. Proefstation voor
de groenteteelt in de volle
grond. Alkmaar

Mededelingen van het Proefstation
voor de groenteteelt in de
volle grond in Nederland

Mededelingen van de Rijksfaculteit
Landbouwwetenschappen te Gent.
Ghent

Mededelingen. Rijksproefstation
voor zaadcontrole. Wageningen

Mededelingen van de Stichting
voor bodemkartering.
Bodemjundige Studies. Wageningen

Mededelingen der Veeartsenij-
school van de Rijksuniversiteit
te Gent. Gent

Mededelingen. Stichting voor
Plantenveredeling. Wageningen

Medecine. Paris

Medecine d'Afrique noire. Dakar

Medecine et hygiene. Geveve

Medical Annals of the District
of Columbia. Washington

Medical Bulletin. Standard Oil
Company. New Jersey

Medical Bulletin of the US Army
Europe.

Medical Clinics of North America.
Philadelphia

Medical History. London

Medical Journal of Australia.
Sydney

Medical Journal of Chulalongkorn
Hospital Medical School.
Thailand

Medical Journal of the Egyptian
Armed Forces. Cairo

Medical Journal of Kagoshima
University.

Medical Journal of Malaya.
Singapore

Medical Journal of Osaka University
Osaka

Medical Officer. London

Medical Press. London

Medical Proceedings. Johannesburg

Medical Radiography and Photography
Rochester

Medical Services Journal, Canada
Ottawa

Medical Technicians Bulletin.
Washington

Medical Times. New York, etc.

Medical World. London

Medicamenta. Madrid

Medicina. Madrid

Medicina. Mexico

Medicina y Cirurgia. Bogota

Medicina y cirurgia de guerra.
Madrid

Medicina, cirurgia, farmacia.
Rio de Janeiro

Medicina clinica. Barcelona

Medicina colonial. Madrid

Medicina contemporânea. Lisboa

Medicina española. Valencia
Medicina Experimentalis.
New York

Medicina internă. Bucuresti

Medicina del lavoro. Milano

Medicina tropical. Madrid

Medicine, analytical Reviews of
general medicine, neurology and
pediatrics. Baltimore

Medicine and Biology. Japan

Medicine Illustrated. London

Medicine and Laboratory Progress
Cairo

Medicinski arhiv. Sarajevo

Medicinski glasnik. Beograd

Medicinski pregled. Novi Sad

Mediterranea. Paris

Mediterranée. Gap

Meditsinskaya parazitologiya i
parazitarnye bolezni. Moskva

Meditsinskaya sestra. Moskva

Meditsinskii referativnii zhurnal.
Moskva. Razdel II.

Medizin und Ernahrung. Lochham-
München

Medizinische Klinik. Berlin, Wien

Medizinische Monatsschrift.
Stuttgart

Medizinische Welt. Stuttgart

Medlemsblad for den danske
Dyrlaegeforening. Kjøbenhavn

Medlemsblad for den norske
Veterinaerforening. Norway

Medycyna doswiadczaina i
mikrobiologia. Warszawa

Medycyna weterynaryjna. Warszawa

Megyei es varosi statisztikai
ertesito. Budapest

Meieriposten. Stenkjaer, etc.

Meijeritieteellinen aikakauskirja.
Helsinki

Meiji-Gakuin Ronso. Tokyo

Mejeri-tidskrift för Finlands
svenskbygd. Helsinki
Mekhanizatsiya i elektrikatsiya
sotsialisticheskgo sel'skogo
khozyaistra. Leningrad

Meldinger fra Norges landbruk-
shøiskole. Kristiania (Oslo).

Meldinger. Statens Forsøksgard
Kvithamar. Stjørdal

Meldinger fra Statens frøkontroll
i As. Norway

Meldinger fra Statens plantevern.
Oslo

Meldinger fra SSR. Tjaereviken-Straumsgrend

Melhoramento. Estudos da Estação de melhoramento de plantas. Elvas

Meliorace. Praha

Memoir. Botanical Survey of South Africa. Pretoria

Memoire del museo civico di storia naturale di Verona

Mémoires de l'Académie de chirurgie. Paris

Mémoires. Academie r. des sciences coloniales. Classe des sciences naturelles et medicales. Bruxelles

Mémoires de l'Institut francais d'Afrique noire. Paris, Dakar

Mémoires de l'Institut r. des sciences naturelles de Belgique. Bruxelles

Memoires de l'Institut scientifique de Madagascar. Tananarive.

Serie A. Biologie animale

Mémoires du Museum nationale d'histoire naturelle. Paris

Mémoires de la Société botanique de France. Paris

Memoires de la Société neuchâteloise des sciences naturelles. Neuchatel

Mémoires de la Société des sciences naturelles et physiques du Maroc. Rabat and Paris

Mémoires de la Société vaudoise des sciences naturelles. Lausanne

Mémoires de l'Université de Neuchatel. Neuchatel

Memoirs. Cambridge University School of Agriculture. Cambridge

Memoirs of the College of Agriculture, Kyoto University. Kyoto

Memoirs of the College of Agriculture, National Taiwan University. Taipei

Memoirs of the College of Medicine, Natural Taiwan University. Taipei

Memoirs of the College of Science, Kyoto University. Kyoto

Series B. Biology

Memoirs. Cornell University Agricultural Experiment Station. Ithaca, NY

Memoirs of Ehime University. Matsuyama

Section 6. Agriculture

Memoirs of the Entomological Society of Canada. Ottawa

Memoirs of the Entomological Society of India. New Delhi

Memoirs of the Entomological Society of Southern Africa. Pretoria

Memoirs of the Faculty of Agriculture, Hokkaido University Sapporo

Memoirs of the Faculty of Agriculture, Kagawa University.

Memoirs of the Faculty of Agriculture, Kagoshima University. Kagoshima

Memoirs of the Faculty of Agriculture, Miyazaki University. Miyazaki

Memoirs of the Faculty of Science. Kyushu University. Fukuoka

Memoirs of the Hyogo University of Agriculture. Sasayama

Memoirs of the Indian Botanical Society. Bangalore

Memoirs of the Indian Museum. Calcutta

Memoirs of the Institute for Protein Research, Osaka University. Osaka

Memoirs of the Queensland Museum. Brisbane

Memoirs of the Research Institute for Food Science, Kyoto University. Kyoto

Memoirs of the Torrey Botanical Club. New York

Memoranda Societatis pro fauna
et flora fennica. Helsingforsiae

Memoranda. West African Maize
Research Unit. Ibadan

Memorandum. Federal Department of
Agricultural Research, Moor
Plantation. Ibadan

Memorandum. Landøkonomiske
Driftsbureau. København

Memorandum. Norges Landbrukshøg-
skole, Institut for Drifts-
laere og Landbruksøkonomi.
Vollebekk

Memoria del Instituto de
Investigaciones Agropecuarias.
Chile

Memorias de la Asociación Latinoamericana de Producción Animal.
Mexico

Memorias. Asociacion de tecnicos
azucareros de Cuba. Havana

Memórias e estudos do Museu
zoológico da Universidade de
Coimbra

Memórias do Instituto Butantan
São Paulo

Memórias do Instituto Oswaldo
Cruz. Rio de Janeiro

Memorias da Junta de Investigações
do Ultramar. Lisboa

Memorias y revista de la
Academia nacional de ciencias.
Mexico

Memórias. Serviços geológicos,
Portugal. Lisboa

Memorias de Sociedade broteriana.
Coimbra

Memorias de la Sociedad de
ciencias naturales 'La Salle'.
Caracas

Memorias de la Sociedad cubana de
historia natural "Felipe Poey"
Habana

Memorie dell'Istituto italiano
di idrobiologia Bott. Marco
de Marchi. Milano

Memorie della Società entomologica
italiana. Genova

Menara perkebunan. Djakarta

Mens en maatschappij. Amsterdam

Mensajero forestal. Durango

Merkblätter. Biologische
Bundesanstalt für Land- und
Forstwirtschaft in Braunschweig.
Braunschweig

Merkblätter. Institute für
Forstwissenschaften, Eberswalde.
Eberswalde

Mesopotamia: the Journal of
Agriculture and Forestry
Research, Mosul, Iraq

Metabolism: Clinical and
Experimental. Baltimore

Meteorological Bibliography.

Meteorological Bulletin.

Meteorological and Geoastro-
physical Abstracts. Lancaster, Pa

Meteorological Magazine. London

Meteorological Research Note.
US Army Electronic Research
Development Activity

Meteorologische Rundschau. Berlin

Meteorologiya i gidrologiya.
Moskva

Methods of Biochemical Analysis.
New York

Methods in Enzymology. New York,
London

Methods in Medical Research.
Chicago

Metra. Paris

Metsanduse Teadusliku Uurimise
Laboratoorium Metsanduslikud
Uurimused, Tallinn, Estonia

Metsätaloudellinen aikakauslehti.
Helsinki

México forestal. México

México y sus bosques. México

Mezdunarodno selskostopansko
spisanie. Sofija

Mezhdunarodnyĭ sel'skokhozyaist-
vennyĭ zhurnal. Sofiya, Moskva

Mezogazdasági Gépészet és
Épitészet. Budapest

Mezogazdasági Könyvtárosok
Tájékoztatója. Budapest

Michigan Botanist. Ann Arbor

Michigan Florist

Michigan State University
Veterinarian. East Lansing

Microbial Genetics Bulletin.
Cold Spring Harbor

Microbiologia española. Madrid

Microbiologia, parazitologia,
epidemiologia. Bucuresti

Microbiology Abstracts.
Sect. A and B. London

Microbiology (USSR). Washington

Microchemical Journal.
New York, London

Microentomology. Stanford
University. Palo Alto

Microscope. London

Midland Bank Review. London

Mie Medical Journal. Tsu-City

Miedzynarodowe Czaspoismo
Rolnicze. Warszawa

Mikologiya i Fitopatologiya.
Leningrad

Mikrobiologichnii Zhurnal. USSR

Mikrobiologiya. Moskva

Mikrobiolohichnyi Zhurnal. Kiev.

Mikrobiyoloji dergisi. Ankara

Mikrochimica acta. Vienna
Mikroelementy. Moskva

Mikroelementy i estestvennaya
radioaktivnost pochv, rostovskii
gosudarstvennyi. Universitet
materialy 3-go mezhvuzovskogo
soveshchaniya. Rostov-na-Domu

Mikroelementy i Produktivnost'
Rastenii. Riga

Mikroelementy zhivotnovodstva
rastenievodstve akademii nauk
kirgizkoi SSR

Mikrokosmos. Stuttgart

Mikroorganizmy i rasteniya trudy
instituta mikrobiologii.
Akademii nauk latviiskoi SSR.
Riga

Mikroorganizmy i Sreda. Riga

Mikroskopie. Wien, etc.

Milbank Memorial Fund Quarterly
Bulletin. New York

Milchwirtschaftliche Berichte aus
den Bundesanstalten Wolfpassing
und Rotholz. Wolfpassing

Milchwissenschaft. Nürnberg

Milk Board Journal. New South
Wales. Sydney

Milk Dealer. Milwaukee

Milk Industry. London

Milk Producer. Johannesburg

Milk Producer. Thames Ditton

Milk Recording Research Results
Pretoria

Military Medicine. Washington

Military Surgeon. Washington

Mimeo. Co-operative Extension
Service, Purdue University
Department of Botany and Plant
Pathology. Lafayette

Mimeograph. Brazzaville Haut
Commissariat. Brazzaville

Mimeograph Series. Georgia
Agricultural Experiment Station.
University of Georgia College of
Agriculture. NS Athens

Mimeographed Series. Institute of
Development Studies, University
of Sussex, Brighton

Mineographed Publication of the
Commonwealth Bureau of
Pastures and Field Crops

Mineral Magazine and Journal of
the Mineralogical Society.
London

Mineralogical Abstracts. London

Minerva chirurgica. Torino

Minerva dermatologica. Torino

Minerva Dietologica. Turin, Milan

Minerva farmaceutica. Torino

Minerva medica. Roma, etc.

Minerva medicolegale. Torino

Minerva paediatrica. Torino

Minerva urologica. Torino

Ministry of Agriculture.
Directorage of Agriculture
Research. Saigon

Minnesota Forestry Research Notes.
Minnesota University School of
Forestry. St. Paul

Minnesota Horticulturist.
Minneapolis

Minnesota Medicine. St. Paul

Minnesota Veterinarian

Minuten Hadeln. Institute for
vector-borne disease control
and research. Thailand

Mirovaia Ekonomika i Mezdunaradnye
Otnosheniia. Moskva

Miscelanea. Facultad de Agronomia
y Zootecnia Universidad
Nacional de Tucumán, Tucumán

Miscelanea zoologica. Spain

Miscelaneas forestals. Administracion Nacional de Bosques.
Buenos Aires

Miscellaneous Information. Tokyo
University Forests. Hongo

Miscellaneous Paper. Agricultural
Economics Research, University
of Sydney. Sydney

Miscellaneous Papers. Pacific
Southwestern Forest and Range
Experiment Station. Berkeley

Miscellaneous Publications.
Commonwealth Mycological
Institute, Kew. London

Miscellaneous Publications.
Economic Research Service,
US Department of Agriculture,
Washington

Miscellaneous Publications.
Edinburgh and East of Scotland
College of Agriculture.
Edinburgh

Miscellaneous Publications of the
Entomological Society of
America. Washington

Miscellaneous Publications of
Fertilizer Research Division.
National Institute of Agricul-
tural Sciences. Tokyo

Miscellaneous Publications. Maine
Agricultural Experiment Station.
Orono

Miscellaneous Publications of the
Maryland Agricultural Experiment
Station

Miscellaneous Publications of the
Museum of Zoology, University
of Michigan. Ann Arbor

Miscellaneous Publications.
Oklahoma Agricultural Experiment
Station. Stillwater

Miscellaneous Publications of the
Pea Growing Research
Organization

Miscellaneous Publications. Rhode
Island Agricultural Experiment
Station. Kingston

Miscellaneous Publication.
School of Agriculture, Aberdeen

Miscellaneous Publications. Texas
Agricultural Experiment Station.
Austin, College Station, Texas

Miscellaneous Publications.
US Department of Agriculture,
Washington

Miscellaneous Publications.
University of Maine. Orono

Miscellaneous Reports of the
Research Institute for Natural
Resources. Tokyo

Miscellaneous Series. New Zealand
Geographical Society.
Christchurch

Miscellaneous Studies. Department
of Agricultural Economics,
University of Reading

Mises au Point de Chemie
Analytique, Organique,
Pharmaceutique et Bromatologique
Paris

Missels zuivelbereiding en
-handel. Doetinchem

Mississippi Doctor. Booneville

Mississippi Farm Research,
State College

Mita Gakkai Zasshi. Tokyo

Mitschruin Bewegung

Mitschurin-Zirkel

Mitteilungen der Arbeitsgemeinschaft zur Förderung des Futterbaues. Zurich.

Mitteilungen der Biologischen Bundesanstalt für Land- u. Forstwirtschaft. Berlin

Mitteilungen der Bundesforschungsanstalt für Forst- und Holzwirtschaft. Berlin

Mitteilungen der Deutschen dendrologischen Gesellschaft. Bonn-Poppelsdorf

Mitteilungen der Deutschen Gesellschaft für Holzforschung. Berlin

Mitteilungen der Deutschen Bodenkundlichen Gesellschaft. Gottingen

Mitteilungen der Deutschen entomologischen Gesellschaft. Berlin

Mitteilungen der Deutschen Landwirtschafts-gesellschaft. Berlin, Frankfurt

Mitteilungen des Deutschen Wetterdienstes. Bad Kissingen

Mitteilungen aus der Forstlichen Bundesversuchsanstalt, Mariabrunn. Wien

Mitteilungen aus dem Gebiet der Lebensmitteluntersuchung und -Hygiene. Bern

Mitteilungen aus dem Hamburgischen zoologischen Museum und Institut. Hamburg

Mitteilungen der Hessischen Landesforverwaltung, Frankfurt

Mitteilungen der Hoheren Bundeslehrund Versuchsanstalten fur Wein-, Obst- und Gartenbau, Wien-Klosterneuburg und fur Bienenkunde, Wien Grinzing

   Ser. A. Rebe und Wein
   Ser. B. Obst und Garten

Mitteilungen Klosterneuburg. Wien

Mitteilungen für den Landbau (BASF)

Mitteilungen des Naturwissenschaftlichen Museums der Stadt Aschaffenburg. Aschaffenburg

Mitteilungen des Naturwissenschaftlichen Vereins für Steiermark. Graz

Mitteilungen Obst und Garten Klosterneuburg

Mitteilungen des obstvauversuchsringes des Alten Landes

Mitteilungen der Österreichischen Bodenkundlichen Gesellschaft

Mitteilungen der Österreichischen Geographischen Gesellschaft. Wien

Mitteilungen für die schweizerische Landwirtschaft. Zurich

Mitteilungen der Schweizerischen Anstalt für das forstliche Versuchswesen. Zurich

Mitteilungen des Schweizerischen Braunviehzuchtverbands. Zug

Mitteilungen der Schweizerischen Entomologischen Gesellschaft. Schaffhausen, Bern

Mitteilungen des Schweizerischen Fleckviehzuchtverbandes. Bern

Mitteilungen aus der Staatsforstverwaltung Bayerns. Bayerisches Staatsministerium fur Ernahrung Landwirtschaft und Forsten. Ministerialforstabteilung. Munchen

Mitteilungen des Vereins fur Forstliche Standortskunde und Forstpflanzensucht. Stuttgart

Mitteilungen der Zentralvorstand VdgB. Berlin

Mitteilungen aus dem Zoologischen Museum in Berlin

Mliekárstvo. Bratislava

Mljekarstvo. Zagreb

Modern Asian Studies. Cambridge

Modern Dairy. Gardenvale, Quebec

Modern Farming. Rhodesia

Modern Problems in Pediatrics. Basle

Modern Veterinary Practice. Chicago

Molecular and General Genetics
Germany

Molini d'Italia. Roma

Molkerei- und Käserei-Zeitung.
Hildesheim

Molkereitechnik. Schriftenreihe
des Zentralverbandes Deutscher
Molkereifachleute und
Milchwirtschaftler. Hildesheim

Molkereizeitung. Welt der Milch.
Hildesheim

Molochnaya promyshlennost'. Moskva

Molocnoe i Mjasnoe Skotovodstvo.
Moskva

Monatsbericht der Deutschen
Akademie der Wissenschaften
zu Berlin. Berlin

Monatsberichte uber die
ostereichische Landwirtschaft.
Wien

Monatsberichte des Osterreichi-
schen Instituts für Wirtschaftsforschung, Wien

Monatsberichte, Schweizerische
Gesellschaft für Konjunktur-
forschung (Ausgabe B: Märkte).
Zurich

Monatshefte für Tierheilkunde.
Stuttgart

Monatshefte für Veterinärmedizin.
Leipzig

Monatsschrift für Kinderheilkunde
Leipzig and Wien

Le Monde Agricole, Chambres
d'Agriculture, Paris

Monde des plantes.

Mondo del latte. Milano

Monitore zoologico italiana.
Firenze

Monografie parazytologiczne.
Poland

Monograph. American Phytopathological Society

Monograph. American Society of
Agronomy. Wisconsin

Monograph. International Studies
in Economics, Department of
Exonomics and Sociology, Iowa
State University, Ames. Iowa

Monograph. National Institute of
Sciences of India. New Delhi

Monograph Series. World Health
Organization, Geneva

Monographiae Biologicae. Den Haag

Monographiae botanicae. Warszawa

Monographien zur angewandten
Entomologie. Berlin

Monographs of the Society for
Research in Child Development.
Washington

Monsanto Technical Bulletin.
St. Louis

Montes. Madrid

Monthly Analytical Bulletin.
Inter-African Bureau for Soils.
Paris

Monthly Bulletin of Agricultural
Economics and Statistics.
FAO Rome

Monthly Bulletin. UP College of
Agriculture, Los Banos

Monthly Dairy Report. Toronto

Monthly Digest of Statistics.
London

Monthly Economic Bulletin.
Ministry of Agriculture,
Republic of Zambia

Monthly Journal. British Goat
Society. London

Monthly Labor Review.
Washington

Monthly Review of the State Bank
of India. Bombay

Monti e boschi. Firenze

Montpellier Médical. Montpellier

Morris Arboretum Bulletin.
Philadelphia

Mosonmagyaróvári Agrartudományi
Fioskola Kozlemenyei. Budapest

Mosonmagyaróvári mesőgazdasági
akadémia kozleménye. Budapest

Mosquito News. New Brunswick, NJ
Albany, NY

Mouse News Letter. New York,
Edinburgh, etc.

Mouton. Paris

Mulino. Bologna

Münchener medizinische Wochenschrift. München

Münchner Studien zur Sozial- und
Wirtschaftsgeographie,
Wirtschaftsgeographisches
Institut der Universitat München,
München

Munkaugyi szemle. Budapest

Mushi. Fukuoka Entomological
Society. Fukuoka

Mushroom News. Worthing.

Mushroom Science. International
Conference on Scientific Aspects
of Mushroom Growing

Mutation Research

Mycologia. Lancaster, Pa. etc

Mycological Note, East African
Agriculture and Forestry
Research Organisation (EAAFRO).
Kikuyu, Kenya

Mycological Papers. Commonwealth
Mycological Institute. Kew

Mycopathologia et mycologia
applicata. Den Haag

Myforest. Forest Department, Mysore

Mykologisches mitteilungsblatt.
Halle

Mykosen. Berlin

Mykrobiolohichniyi Zhurnal. Kyyiv

Mysore Agricultural Journal.
Bangalore

Mysore Journal of Agricultural
Sciences. Hebbal, Bangalore

N

N.A.A.S. Advisory Papers.
National Agricultural Advisory
Service. London

N.A.A.S. Progress Report.
National Agricultural Advisory
Service. London

N.A.A.S. Quarterly Reviews.
National Agricultural Advisory
Service. London

N.I.B.S. Bulletin of Biological
Research Nippon Institute for
Biological Science. Tokyo

N.L.L. Translations Bulletin.
National Lending Library

N.S.F. Information. Nämnden för
Skoglig Flygbildteknik.
Vällingby

Nachrichtenblatt der Bayerischen
Entomologen. München

Nachrichtenblatt der Biologischen
Zentralanstalt, Braunschweig.
Stuttgart

Nachrichtenblatt für den
Deutschen Pflanzenschutzdienst.
Berlin, etc.

Nachrichtenblatt des Deutschen
Pflanzenschutzdienstes.
Stuttgart

Nachrichten des Naturwissenschaftlichen Museums der Stadt
Aschaffenburg

NacoBrouw jaarboekje. Rotterdam

Nagoya Journal of Medical
Science. Nagoya

Nagoya Medical Journal. Nagoya

Nagpur Agricultural College
Magazine. Nagpur

Nahrung. Germany

Näringsforskning. Uppsala

Narodi Azii i Afriki. Moskva

Narodnostopanski Arhiv. Svistov

Náš chov. Praha

National Agricultural Chemicals
Association News and Pesticide
Review. Washington

National Cactus and Succulent
Journal. Bradford

National Farmers' Union
Information Service. London

National Institute of Animal
Health Quarterly. Tokyo

National Medical Journal of China.
Shanghai

National Provincial Bank Review.
London

National Research Council
Review. Ottawa

National Wool Grower. Salt Lake
City

Natur und Museum. Senckenbergische
naturforschende Gesellschaft.
Frankfurt a.M.

Natura. București

Natural History. New York

Natural History Notes (Natural
History Society of Jamaica)

Natural History Papers. National
Museum of Canada. Ottawa

Natural Resources Journal.
New Mexico

Naturalia. Sociedade portuguesa
de ciencias naturais. Lisboa

Naturalist. Hull, London

Naturaliste malgache.
Tananarive-Tsimbazaza

Nature. London

Nature. Paris

Nature Conservancy Handbook. UK

Nature Conservancy Progress. UK

Nature and Resources. UNESCO

Naturwissenschaften. Berlin

Naturwissenschaftliche Rundschau.
Stuttgart

Natuurhistorisch maanblad.
Naastricht

Natuurwetenschappelijk tijdschrift.
Antwerpen

Natuurwetenschappelijke Studiekring
voor Suriname en de Nederlandse
Antillen.

Nauchnaya Konferentsiya po Yader-
noi Meteorologii. Obninsk

Nauchni trudove. Bulgarska
akademiya na naukite, otdelenie
za selskostopanski nauki.
Bulgaria

Nauchni trudove. Nauchnoizsle-
dovatelski Institut Proektokonstru
Mehanizacija Traktorno i
Selskostopanska masinostroene.
Sofija

Nauchni trudove. Tsentralen
Veterinaren institut za
Zarazni i Parazitni Bolesti.
Bulgaria

Nauchni trudove. Vissh Institut
po Khranitelna i Vkusova
Promishlennost. Plovdiv

Nauchni trudove. Vissh
lesotekhnicheski institut.
Sofiya

Nauchni trudove. Vissh
selskostopanski institut
'Georgi Dimitrov'. Sofiya

Zootekhnicheski Fakultet
Agronomicheski Fakultet

Nauchni trudove. Vissh
selskostopanski institut
'Vasil Kolarov'. Plovdiv

Nauchni trudove. Vissh
veterinarnomeditsinski institut
'Prof. G. Pavlov'. Sofiya

Nauchno-tekhnicheskaya
informatsiya po melioratsii i
gidrotekhniki

Nauchnȳe dokladȳ vȳsshei shkolȳ.
Moskva

Biologischeskie nauki

Nauchnȳe trudȳ Altaiskogo
Nauchno-Issledovatel'skogo
Instituta Sel'skogo Khozyaistva.

Nauchnȳe trudȳ Kharkovskogo
sel'sko- Khozyaistvennogo
Instituta. Kharhov.

Nauchnȳe trudȳ Leningradskaya
lesotekhnicheskaya akademiya
imeni S.M. Kirova. Leningrad

Nauchnȳe trudȳ Litovskoi sel'skokhozyaistvennoi akademii.Velnyus

Nauchnȳe trudȳ Severo-Zapadnyi
Nauchno-issledovatels'kii Insti-
tut Sel'skogo Khozyaistva.
Leningrad

Nauchnȳe trudȳ Tashkentskogo
Universiteta. Tashkent

Nauchnye trudy. Ukrainski
Nauchnoissledovatelski
institut eksperimentalnoi
veterinarii.

Nauchnye trudy Ukrainskog Nauchno-
Issledovatel'skogo Instituta
Lesnogo Khozyaistva i Agroleso-
melioratsii.

Nauchnye trudy Ukrainskogo
Nauchnoissledovatel'skogo
instituta Pochvovedeniya.
Kiev

Nauchnye zapiski. Belotserkovski
selskokhozyaistvenni institut.
USSR

Nauchnye zapiski. Voronezheskogo
Lesotekhozyaistvennogo
instituta. USSR

Naucni Trudove. Lesotehnioeski
Institut, Sofia (Serija
Mehanicna Tehnologija na
Dárvesinata

Naucni Trudove Vissija Institute
po Mehanizacija i Elektrifikat-
cija na selskoto stopanstvo.
Ruse

Nauka i Ziznj

Naukovi pratsi. Lovvski
zooveterinarni Institut.

Naukovi pratsi. Ukrainskii
Naukovo-Doslidnii Institut lis.
Gospod. Agrolosomelior. Kiiv

Naukovi zapiski. Cherkaski
derzhavni pedagogichni institut.

Naunyn-Schmiedebergs Archiv für
Experimentelle Pathologie und
Pharmakologie. Leipzig

Naval Stores Review. New Orleans

Nebraska Experimental Station
Quarterly. Lincoln

Nederlands Bosbouwtijdschrift.
Wageningen

Nederlands melk en zuiveltijd-
schrift. Amsterdam (Netherlands
Milk and Dairy Journal)

Nederlands tijdschrift voor
geneeskunde. Amsterdam

Nederlandsch boschbouwtijdschrift.
Wageningen

Neftyanye udobreniya i stimulatory,
akademiya nauk azerbaidzhanskoi
SSR. Otdelenie sel'skokhozyaistvennykh Nauk. Baku

Nehézvegyipari kutató intézet
közleményei. Veszprém

Nematologica. Leiden

Neotropica. Notas zoologicas
sudamericanas. Buenos Aires

Népegészségügy. Budapest

Népszabadság. Budapest

Netherlands Journal of
Agricultural Science. Wageningen

Netherlands Journal of Plant
Pathology. Wageningen

Netherlands Journal of Veterinary
Science (Tijdschrift voor
Diergeneeskunde

Netherlands Nitrogen Technical
Bulletin. The Hague

Neue gesellschaft. Bielefeld

Neue Verpackung. Berlin

Neues Archiv für Niedersachsen.
Göttingen

Neues Jahrbuch für Mineralogie,
Geologie und Paläontologie.
Monatshefte. Stuttgart

Neumologia y cirugia de torax.
Mexico

Neurocirugia. Santiago de Chile

Neuroendocrinology. Basle

Neurologia, neurochirurgia i
psychiatria polska. Warszawa

Neurology. Minneapolis

Neuropsiquiatria. Buenos Aires

Nevada Ranch and Home Review

New Biology. London, New York

New Commonwealth

New England Journal of Medicine
Boston

New Guinea Research Bulletin.
New Guinea Research Unit,
Australian National University.
Canberra

New Hungarian Quarterly. Budapest

New Mexico Business. Albuquerque

New Mexico Extension News

New Middle East. London

New Phytologist. Cambridge, etc.

New Scientist. London

New York's Food and Life
Sciences. Cornell

New York State Journal of Medicine.
New York

New Zealand Agricultural Science

New Zealand Agriculturist. Fielding

New Zealand Commercial Grower.
Wellington

New Zealand Dairy Exporter.
Wellington

New Zealand Engineering. Journal
of the New Zealand Institution
of Engineers. Wellington

New Zealand Entomologist. Nelson

New Zealand Forestry Research
Notes. Forest Research Institute,
New Zealand Forest Service.
Rotorua

New Zealand Gardener. Wellington

New Zealand Geographer.
Christchurch

New Zealand Journal of
Agricultural Research.
Wellington

New Zealand Journal of
Agriculture. Wellington

New Zealand Journal of Botany.
Department of Scientific and
Industrial Research.
Wellington

New Zealand Journal of Dairy
Technology. Palmerston North

New Zealand Journal of Forestry.
Christchurch

New Zealand Journal of Geology
anf Geophysics. Wellington

New Zealand Journal of Science.
Wellington

New Zealand Medical Journal.
Wellington

New Zealand Plants and Gardens.
Royal New Zealand Institute of
Horticulture. Wellington

New Zealand Science Review
Wellington

New Zealand Timber Journal and
Forestry Review. Auckland

New Zealand Veterinary Journal.
Wellington

New Zealand Wheat Review

News in Engineering at the Ohio
State University College of
Engineering. Columbus

Newsletter. Arid Zone Research.
UNESCO

Newsletter on the Common Agricultural Policy, European Economic
Community. Brussels

Newsletter of the FAO International
Rice Commission

Nigerian Agricultural Journal.

Nigerian Field. Journal of the
Nigerian Field Society. London

Nigerian Forestry Information
Bulletin. Ibadan

Nigerian Geographical Journal
Ibadan

Nigerian Journal of Economic and
Social Studies. Ibadan

Nigerian Journal of Science.

Nigerian Trade Journal. Lagos

Niigata Agricultural Science.
Faculty of Agriculture,
Niigata University, Niigata

Niigata Daigaku Keizai Ronshu.
Niigata

Niigata Medical Journal. Niigata

Niigata Norin Kenkyu

Nippon Journal of Angio-Cardiology.
Kyoto

Nogaku kenkyu. Japanese Report
of the Ohara Institute for
Agricultural Biology

Nogyo Gijutsu Kenyusho
(Series H). Tokyo

Nogyo Keizai Kenkyu. Tokyo

Nogyo sogo kenkyu. Tokyo

Nokei Ronso. Sapporo

Nord e Sud. Napoli

Nordisk hygienisk tidskrift.
Lund, etc.

Nordisk jordbruksforskning.
Kristiania, etc.

Nordisk lantbruksekonomisk
tidskrift. Stockholm

Nordisk medicin. Stockholm, etc.

Nordisk mejeritidsskrift.
København

Nordisk Udredningsserie.
Stockholm

Nordisk veterinærmedicin.
Denmark

Norges geologiske undersøgelse.
Kristiania (Oslo)

Norges landbrukshøgskoles
beretning fra foringsforsøkene.
Kristiania

Norges Vel. Oslo

Noringyo Mondai Kenkyu. Osaka

Norrlands Skogsvärdsförbunds
Tidskrift. Stockholm

Norsk entomologisk tidsskrift.
Kristiania (Oslo)

Norsk Geografisk Tidsskrift.
Oslo

Norsk hagetidende. Oslo

Norsk landbruk. Oslo

Norsk pelsdyrblad. Norway

Norsk skogbruk. Oslo

Norsk skogindustri. Oslo

Norske myrselskap

North American Veterinarian.
Evanston, Ill.

North Carolina Medical Journal
Winston-Salem

North Central Regional Extension
Publication Cooperative
Extension Service, Iowa State
University of Science and
Technology, Ames, Iowa

North Dakota Farm Research. Fargo

Northern Gardener. Manchester

Northwest Science. Cheney

Nosokomeiaka hronika. Athenai

Nota pratica. Stazione di
entomologia agraria. Firenze

Nota tecnica. Instituto forestal
Santiago

Notas agronomicas. Estacion
agricola experimental de
Palmira. Republica de Colombia

Notas del Museo de La Plata.
Buenos Aires.

Zoologia

Notas Silvicolas, Administración
Nacional de Bosques, Buenos
Aires

Notas tecnicas. Instituto
nacional de investigaciones
forestales. Secretaria de
Agricultura y Ganaderia. Mexico

Notas Técnico Forestales, Escuela
de Ingenieria Forestal,
Universidad de Chile, Santiago

Notas Tecnológicas Forestales,
Administración Nacional de
Bosques, Buenos Aires

Note de Recherches, Faculté de
Foresterie et Géodésia,
Université Laval, Quebec

Note. Rocky Mountain Forest and
Range Experiment Station.
Fort Collins

Note Technique. Centre Technique
Forestier Tropical. Nogent sur
Marne

Note Technique, Faculté de
Foresterie et Géodésie,
Université Laval, Quebec

Note Technique, Institut de
Reboisement de Tunis. Ariana

Note. United States Forest Service,
Lake States Forest Experiment
Station. St. Paul, Minn

Notes from the Botanical School of
Trinity College. Dublin

Notes et Etudes documentaires.
Paris

Notes. Meteorologie Nationale.
Paris

Notes from the Royal Botanic
Garden, Edinburgh

Noticias agricolas. Servicio
Shell para el agricultar.
Cagua, Aragua

Notizblatt des Hessischen
Landesamtes für Bodenforschung
zu Wiesbaden

Notiziario sulle malattie delle
piante. Milano, Pavia

Notulae entomologicae.
Helsingfors

Notulae naturae. Academy of
Natural Sciences of Philadelphia

Nouvelles des marches agricoles.
Paris

Nova acta Academiae Caesareae
Leopoldino Carolinae germanicae
naturae curiosorum. Halle a.S.

Nova Hedwigia. Zeitschrift für
Kryptogramenkunde. Weinheim

Nova mysl. Praha

Novedades cientificas.
Contribuciones ocasionales del
Museo de historia natural La
Salle. Caracas

Serie zoológica

Novenynemesitesi es Novenytermesztesi Kukako Intezet
Kozlemenyti, Sopronhopaos

Növenytermeles. Budapest

Novenyvedelmi Kutato Intezet
Evkonyve. Annales Instituti
Protectionis Plantarum Hungarici.
Budapest

Novo Vreme. Sofiya

Novos taxa entomológicos.
Lourenço Marques.

Novosti sistematiki vysshikh
rastenii. Novitates systematicae
plantarum non vascularum.
Akademiya Nauk SSSR.
Botanicheskii Institut im. V.L.
Komarova. Moskva, Leningrad

Nowe drogi. Warszawa

Nowe rolnictwo. Warszawa

Nuclear Instruments and Methods.
Amsterdam

Nuclear Science Abstracts.
Oak Ridge, Tenn.

Nucleus. International journal of
cytology and allied subjects.
Calcutta

Nukleonika. Warszawa

Nuova veterinaria. Bologna, Faenza

Nuovi annali d'igiene e
microbiologia. Roma

Nuovi annali. Istituto chimico-
agrario sperimentale di Gorizia.
Gorizia

Nuovo giornale botanico italiano.
Firenze

Nurseryman and Seedsman. London

Nutricion, bromatologia,
toxicologia. Santiago, Chile

Nutritio et dieta. Basel, New York

Nutrition Abstracts and Reviews
Aberdeen

Nutrition. Journal of Dietetics,
Food, Catering and Child
Nutrition. London

Nutrition Reviews. New York

Ny Jord. Oslo

Nyasaland Farmer and Forester.
Zomba

Nytt magasin for botanikk. Oslo

Nytt magasin for zoologi. Oslo

O

O.E.C.D. Agricultural Review. Paris

O.E.C.D. Observer. Paris

Oat Newsletter

Obeche: Journal of the Tree Club.
Ibadan

Obst- und Gartenbaulicher
Beratungsund Informationsdienst
Weihenstephan. Weihenstephan

Obstbau. Stuttgart

Obstetrics and Gynecology. New York

Occasional Papers of the Bernice P. Bishop Museum. Honolulu

Occasional Paper. Bureau of Forestry, Philippines. Manilla

Occasional Papers of the C.C. Adams Center for Ecological Studies. Kalamazoo

Occasional Paper. Centre for Asian Studies, Arizona State University. Tempe

Occasional Paper. Department of Agricultural Economics and Farm Management, Massey University, Palmerston North, N.Z.

Occasional Papers, Farm Economics Branch, Department of Land Economics, Cambridge University. Cambridge

Occasional Paper. Institute for Development Studies, University College, Nairobi

Occasional Papers. Mauritius Sugar Industry Research Institute. Reduit

Occasional Papers of the Museum of Zoology, University of Michigan. Ann Arbor

Occasional Papers. National Museum of Southern Rhodesia. Bulawayo

Occidental Entomologist. Eureka

Ochrana rostlin. Praha

Ochrona Roślin. Poland

Oecologia plantarum. Paris

Offene Welt. Opladen

Öffentliche Gesundheitsdienst. Leipzig. Stuttgart

Office de la Recherche Scientifique et Technique Outre-Mer. Service des Sols. Laboratoire de Chimie. Paris (ORSTOM)

Official Journal, Patents. London

Officieel Orgaan van de Koninklijke Nederlandse Zuivelbond. The Hague

Ohio Florist Association Bulletin.

Ohio Journal of Science. Columbus

Ohio Report on Research and Development in Biology, Agriculture and Home Economics.

Ohio State Medical Journal. Columbus

Oikos. Acta ecologica scandinavica. København

Okajimas folia anatomica japonica. Tokyo

Okhrana prirody na dalnem vostoke Akademiya nauk SSSR. Vladivostok

Oklahoma Current Farm Economics. Stillwater

Oklahoma Farm Research Flashes

Økonomi og Politik. København

Oléagineux. Revue générale des corps gras et dérivés. Paris

Olearia. Revista delle materie grasse. Roma

Olien, vetten en zeep. Bussum

Oncologia. Basel, New York

Onderstepoort Journal of Veterinary Research. Onderstepoort, Pretoria

Opera corcontica. Praha

Ophthalmologica. Basel

Opredeliteli po faune SSSR. Moscow

Optima. Anglo-American Corporation of South Africa. Johannesburg

Opuscula entomologica. Lund

Opuscula zoologica. Instituti zoosystematici Universitatis budapestinensis. Budapest

Orbis. Philadelphia

Orchardist of New Zealand. Wellington

Oregon State Monographs. Studies in Zoology. Corvallis

Organic Mass Spectrometry. London

Organic Reactions. New York

Oriental Geographer. Pakistan

Oriente agropecuario. Venezuela

Orman ve av. Ankara

Ormancilik araştirma enstitusu
dergisi. Ankara

Ormancilik Araştirma Enstitusu
Mahtelif Yayinlar Serisi.
Ankara

Ormancilik Araştirma Enstitusu
Teknik Bülten. Ankara

Ormancilik Araştirma Enstitusu
Yillik Bülten. Ankara

Országos közegészségügyi intézet
muködese. Budapest

Országos meteorológiai intézet
hivatalos kiadvanyai. Budapest

Országos mezögazdasági
minöségvizsgáló intézet evkönyve
Budapest

Országos mezögazdasagi minöségvizsgáló intézet kiadvanyai.
Budapest

Ortopedia e traumatologia dell'-
apparato motore. Roma

Orvosi hétilap es szovjet
orvostudomanyi beszamolo.
Budapest

Oryx. Journal of the Fauna
Preservation Society. London

Oryza. Cuttack

Osaka City Medical Journal. Osaka

Osaka daigaku igaku zasshi. Osaka
/English edition at: Medical
Journal, Osaka University/

Osaka Plant Protection. Osaka
Osmania Agriculturist

Ospedale maggiore. Milano

Österreichisches Bank-Archiv. Wien.

Österreichische botanische
Zeitschrift. Wien

Österreichische milchwirtschaft.
Wien

Österreichische ost-hefte. Wien

Österreichische Zeitschrift fur
Stomatologie. Wien

Österreichische zoologische
Zeitschrift. Wien

Osteuropa. Stuttgart

Osteuropa-Recht. Stuttgart

Osteuropa-Wirtschaft. Stuttgart

Osteuropastudies der Hochschulen
des Landes Hessen. Wiesbaden

Otčetnost i Kontrol selskoto
Stopanstvo. Sofija

Outlook on Agriculture.
Bracknell

Overseas Business Reports
Washington

Ovoshtarstvo. Mesechno spisanie
na Ministerstvoto na
Selskostopanskoto proizvodstvo.
Sofia

Ovtsevodstvo. Moskva

Oxford Economic Papers. Oxford

P

P.A.N.S. Pest Articles and
News Summaries. London

P.I.D.A. Technical Bulletin.
London

P.U.D.O.C. Bulletin. Centre for
Agricultural Publications and
Documentation. Wageningen

Pacific Affairs. Vancouver

Pacific Insects. Honolulu

Pacific Naturalist. Beaudette
Foundation for Biological
Research. Solvang

Pacific Science. Honolulu

Pacific Viewpoint. Wellington

Package Engineering. Chicago, Ill

Packaging News. London

Packaging Review. London

Padiatrie und Padologie. Vienna

Pakistan Cottons. Karachi

Pakistan Development Review.
Karachi

Pakistan Economic Journal

Pakistan Geographical Review.
Lahore

Pakistan Journal of Animal Sciences

Pakistan Journal of Biological and
Agricultural Sciences. Dacca

Pakistan Journal of Forestry
Abbottabad

Pakistan Journal of Health.
Madras. Karachi

Pakistan Journal of Medical
Research. Karachi

Pakistan Journal of Science.
Lahore

Pakistan Journal of Scientific
and Industrial Research. Karachi

Pakistan Journal of Scientific
Research. Lahore

Pakistan Journal of Soil Science.
Dacca

Pakistan Journal of Veterinary
Science.

Pakistan Quarterly. Karachi

Pakistan Review of Agriculture.
Karachi

Palaeobotanist. Birbal Sahni
Institute of Palaeobotany.
Luchnow

Palestine Journal of Botany.
Jerusalem Series. Jerusalem

Palestine Journal of Botany.
Rehovot Series. Rehovot

Pamietnik Pulawski. Bydgoszcz

Pamphlet. Colorado State
University Experiment Station.
Fort Collins

Pamphlet. Department of
Agriculture, Tanzania

Pamphlet. Forestry Research and
Education Project, Forests
Department. Khartoum

Pamphlet. Volcani Institute of
Agricultural Research. Rehovot

Pan-Pacific Entomologist.
San Francisco

Papéis avulsos do Departamento
de Zoologia. São Paulo

Paper Trade Journal. New York
and Chicago

Paperi ja Puu. Helsinki

Papers. Commonwealth Forestry
Conference

Papers from the Department of
Botany, University of Queensland.
Brisbane

Papers. Department of Entomology,
University of Queensland.
Brisbane

Papers. Economic Social Research
Institute. Dublin

Papers. Faculty of Veterinary
Science, University of
Queensland

Papers. Laboratory of Tree-Ring
Research. University of Arizona

Papers from the Michigan Academy
of Sciences, Arts and Letters.
New York

Papers. National Symposium on
Water Resources. Use and
Management. Canberra

Papers of the Northeastern Forestry
Experiment Station

Papers and Proceedings of the
Royal Society of Tasmania.
Hobart

Papers. Rocky Mountain Forest and
Range Experiment Station.
Fort Collins

Papers. Technical Meeting of the
International Union for
Conservation of Nature and
Natural Resources. (IUCN). Morges

Papers. United States Geological
Survey. Water Supply. Washington

Papers. University of Queensland.
Department of Agriculture.
St. Lucia

Papier. Darmstadt

Papua and New Guinea Agricultural
Journal. Port Moresby

Papua and New Guinea Medical
Journal. Port Moresby

Parallèlles. Genève

Parasitica. Gembloux

Parasitologische Schriftenreihe. Jena

Parasitology. Cambridge

Parassitologia. Roma

Parazitologicheskii sbornik. Moskva

Park Administration

Parks and Sports Grounds. London

Pártélet. Budapest

Pastoral Review and Grazier's Record. Melbourne, Sydney

Pathologia Europaea. Brussels

Pathologia et microbiologia. Basel and New York

Pathologia veterinaria. Switzerland

Pathologie et biologie. Paris

Patologia polska. Warszawa

Patre. Revue mensuelle de l'elevage ovin.

Paysans. Paris

Pchelovodstvo. Moskva

Peat Abstracts. Dublin

Pecuaria. Anais dos Servicos de veterinaria e industria animal. Angola. Loanda

Pediatria. Barcelona

Pediatria. Naples

Pediatria polska. Warszawa

Pediatria pratica. São Paulo

Pediatric Clinics of North America. Philadelphia, London

Pediatric Research.

Pediatrics. New York

Pédiatrie. Paris

Pediatriya. Moskva

Pediatriya, Akusherstvo i Ginekologiya. Kiev

Pedobiologia. Jena

Pédologie. Société belge de pédologie. Gand

Pedologist. Tokyo

Pemberitaan. Balai besar penjelidikan pertanian. Bogor

Pendik Veteriner Kontrol ve Araştirma Enstitüsü Dergisi. Turkey

Penggemar alam. Bogor

Pengumuman Lembaga Penelitian Kehutanan (Communication Forest Research Institute). Bogor

Pennsylvania Medical Journal. Pittsburg

Pénzügyi Szemle. Budapest

Perfumery and Essential Oil Record. London

Persoonia. A mycological journal. Leiden

Perspectives in Biology and Medicine. Chicago

Perspectives socialistes. Paris

Pesquisa Agropecuária Brasileira. Brazil

Pest Control. Painesville, Ohio

Pest Infestation Research. Report of the Pest Infestation Research Board, D.S.I.R. London

Pesticide Information. London

Pesticides Monitoring Journal. Washington

Pesticide Progress. Ottawa

Pesticide Review. Washington

Pesticides Documentation Bulletin Washington

Pesticides Quarterly Supplement. Canberra

Petermanns geographische Mitteilungen. Gotha

Pflanzenarzt. Wien

Pflanzenschutz. Munchen

Pflanzenschutzberichte. Wien

Pflanzenschutz-Merkblatt. Germany

Pflanzenschutz Nachrichten Bayer. Leverkusen

Pflügers Archiv - European Journal of Physiology. Berlin

Pharmaceutical Bulletin. Tokyo

Pharmaceutical Journal. London

Pharmaceutisch weekblad voor
Nederland. Amsterdam

Pharmacological Reviews.
Baltimore

Pharmazeutische Zentralhalle für
Deutschland. Dresden

Pharmazie. Berlin

Philippine Abstracts. Manila

Philippine Agriculturist.
Los Baños

Philippine economic Journal,
Manila

Philippine Farmers' Journal.
Manila

Philippine Forests. Manila

Philippine Journal of Agriculture.
Manila

Philippine Journal of Animal
Industry. Manila

Philippine Journal of Forestry.
Manila

Philippine Journal of Nutrition.
Manila

Philippine Journal of Plant
Industry. Manila

Philippine Journal of Public
Administration. Manila

Philippine Journal of Science.
Manila

Philippine Journal of Veterinary
Medicine. Manila

(Philippine) Lumberman, Manila

Philippine Review of Business and
Economics. Diliman

Philippine Sugar Institute
Quarterly. Manila

Philosophical Transactions of the
Royal Society. London

Series B

Phosphorsäure. Berlin, etc

Photogrammetric Engineering
Washington, etc.

Photosynthetica. Czechoslovakia

Physics in Medicine and Biology
London

Physiologia bohemoslovenica. Praha

Physiologia plantarum. København

Physiological Reviews. Baltimore

Physiological Zoölogy. Chicago

Physiologie des plantes tropicales
cultivees

Physiologie Vegetale

Physiology and Behaviour. Oxford

Physis. Revista de la Sociedad
argentina de ciencias naturales.
Buenos Aires

Phytiatrie-phytopharmacie. Paris

Phytochemistry. Oxford

Phytologia. New York

Phytoma. Revue de phytomédecine
appliquée. Paris

Phytomorphology. Delhi

Phyton. Buenos Aires

Phyton. Graz
Phyton. Horn

Phytopathologia mediterranea.
Bologna

Phytopathological Classics.
American Phytopathological
Society. Lancaster, Pa

Phytopathological Papers.
Commonwealth Mycological
Institute. Kew

Phytopathologische Zeitschrift.
Berlin

Phytopathology. American Phytopathological Society.
Worcester, Mass.

Phytoprotection. Ministere de
l'Agriculture et de la
Colonisation du Quebec. Service
d'information et de Recherches.
Montreal

Piackutatas. Budapest

Pienpuualan Toimikunan Julkaisu,
Helsinki

Pig Breeders' Gazette. London,etc.

Pig Farming. Ipswich

Pishchevaya Promyshlennost'. Kiev

Pitanie i Udrobenie Rastenii. Kiev

Plan. Zürich

Planning, P.E.P. London

Planovane hospodarstvi. Praha

Planovo stopanstvo i statistika
Sofiya

Planovoe khozyaistvo. Moskva

Plant Breeding Abstracts.
Cambridge.

Plant and Cell Physiology. Tokyo

Plant Disease Leaflet. New South
Wales Department of Agriculture,
Division of Science Services,
Biology Branch. Sydney

Plant Disease Reporter. Washington

Plant Disease Survey. Annual
Report. New South Wales
Department of Agriculture,
Division of Science Services,
Biology Branch. Sydney

Plant Foods for Human Nutrition.
Oxford

Plant Food Review. New Delhi

Plant Food Review. Washington

Plant Industry Digest. Manila

Plant Industry Series. Chinese-
American Joint Committee on
Rural Reconstruction. Taipei

Plant Introduction Newsletter.
FAO Rome

Plant Introduction Review.
Canberra

Plant Inventory. US Department of
Agriculture. Washington

Plant Pathology. London

Plant Pathology Circular. Division
of Plant Industry, Florida
Department of Agriculture.

Plant Physiology. Lancaster, Pa.

Plant Protection Abstracts.
Makhteshim Chemical Works.
Beer-Sheva

Plant Protection Bulletin.
FAO Rome

Plant Protection Bulletin.
New Delhi

Plant Regulatory Announcements,
Plant Pest Control Division
/and/ Plant Quarantine Division,
US Department of Agriculture,
Washington

Plant and Soil. The Hague

Plant Varieties and Seed Gazette

Planta. Archiv für wissenschaftliche Botanik. Berlin

Planta medica. Stuttgart

Plantation Health. Honolulu

Planter. Kuala Lumpur

Planters' Bulletin. Rubber
Research Institute of Malaya.
Kuala Lumpur

Planters' Chronicle. Coimbatore

Plantesygdomme i Danmark.
Arsoversigt samlet ved statens
plantepatologiske forsøg.
København

Plants and Gardens. Brooklyn
Botanic Garden. New York

Plodorodie Pochv Karelii,
Akademiya Nauk SSSR. Karel' skii
Filial. Moskva

Pochvovedenie. Leningrad

Pochvoznanie i Agrokhimiya. Sofia

Pochvy Dolin Rek Leny i Aldana.
Yakutsk

Polar Record. A journal of Arctic
and Antarctic Research.
Cambridge

Policlinico. Roma.

Sezione pratica

Polish Scientific Periodicals.
Current Contents

Polish sociological Bulletin.
Warsaw

Polish Technical Abstracts.
Warszawa

Politicka Ekonomie. Praha

Politicka misao. Zagreb

Politique. Paris

Politique etrangere. Paris

Pollen et spores. Paris

Polnohospodárstvo. Bratislava

Polski przegląd chirurgiczny.
Krakow, Warszawa

Polski przegląd radiologiczny.
Warszawa

Polski tygodnik lekarski i
wiadomości lekarskie.
Warszawa

Polskie archiwum hydrobiologii.
Warszawa

Polskie archiwum medycyny
wewnetrznej. Warszawa

Polskie archiwum weterynaryjne.
Warszawa

Pomme de terre française. Lille

Pomologie française. Lyon

Poona Agricultural College
Magazine. Poona

Popularne monografie zoologiczne.
Poland

Poljoprivredna Znanstvena Smotra.
Poljoprivrednog Fakulteta,
Sveučilista u Zagrebu. Yugoslavia.

Population Bulletin. Washington

Population Studies, Cambridge UK

Populier. Wageningen

Portugal medico. Arquivos
portugueses de medicina. Porto

Portugaliae acta biologica.
Lisboa

   Serie A
   Serie B

Postępy biochemii. Warszawa

Postępy higieny i medycyny
doświadczalnej. Warszawa

Postepy nauk rolniczych. Warszawa

Post-Graduate Medical Journal.
London

Postgraduate Medicine. Milwaukee

Power Farming in Australia and
New Zealand. Sydney

Potash Review. International
Potash Institute. Berne

Potash and Tropical Agriculture
Amsterdam

Potasse. Mulhouse

Potassium Symposium. International
Potash Institute. Berne

Potato Journal. Christchurch

Potato News from the Netherlands.
's Gravenhage

Poultry Digest, USA

Poultry Industry. Godalming, etc.

Poultry Review

Poultry Science. Ithaca, NY

Poumon et le coeur. Paris

Poznańskie Towarzystwo Przyjaciół
Nauk, Wydział Matenotyczno-
Przyrodniczy, Prace Komisji Nauk
Rolniczych i Lesnych.

Poznańskie Towarzystwo Przyjaciół
Nauk, Wydział Matematyezno-
Przyrodniezy, Prace Komisji geo-
grapiezno-geologreznej. Poznań

Práce Brněnské základny, Ceskoslovenské akademie věd. Brno

Prace instytut technologii drewna.
Poznan

Prace instytuti techniki bodowlan-
ej. Warszawa

Prace instytutu badawczy leśnict-
wa. Warszawa

Prace instytutu przemyslu mleczarskiego. Warszawa

Prace instytutu sadownictwa w
Skierniewicach. Warszawa

Prace Komisji biologicznej.
Poznańskie towarzystwo
przyjaciól nauk. Poznań

Prace Komisji nauk rolniczych i
leśnych. Poznańskie towarzystwo
przyjaciól nauk. Poznań

Prace Materiały Naukowe Instytut
Matki i Dziecka. Warsaw

Prace a Mzda. Praha

Prace naukowe Instytutu ochrony
roslin. Warszawa

Práce Výzkumných ustavu
lesnickych. Praha

Práce Vyzkummého Ústavu Lesniho
Hospodárství a Myslivosti
(VULHM), Zbraslav-Strnady

Prace Výzkumného Ústavu
Narodohospodárského Planováni.
Praha

Prace Wroclawskiego towarzystwa
naukowego. Wroclaw

Practicing Veterinarian. USA

Practitioner. London

Praktický lékař. Praha

Praktische Tierarzt. Hannover

Pratsi Botanichnoho sadu.
Akademiya nauk ukrayinskoyi RSR
Kiyiv

Prazis. Switzerland

Pregled naucnotehnickih radova i
informacija, Zlavod za tehnolo-
giju drveta. Sarajevo

Přehled lesnické a myslivecké
literatury. Praha
Prensa médica argentina.
Buenos Aires

Prensa médica mexicana. México

Prensa pediátrica. Buenos Aires

Preprints. American Chemical
Society. Division of Fuel
Chemistry. Washington

Preprints. American Wood Preservers
Association. Washington

Preslia. Časopis československé
botanické společnosti. Praha

Presse médicale. Paris

Previdenza sociale nell'
Agricoltura. Roma

Prikladnaya Biokhimiya i
Mikrobiologiya. Moscow

Primenie mikroelementov, polimerov
i radioaktivnykh izotop v
sel'skom khozyaistve, ukrainsk-
aya akademiya sel'skokhozyaistvennykh nauk, trudy koordinat-
sionnogo soveshchaniya. Kiev.

Principes. Palm Society. Miami

Přirodovědecky sbornik
Ostravského kraje. Opava

Priroda. Moskva, Petrograd

Probleme agricole. București

Probleme economice. București

Probleme des Friedens und des
Sozialismus. Berlin

Probleme de Economie Agrara.
Bucuresti

Probleme de parazitologie
veterinara. Institutul de
pathologie si igiena animala.
București

Probleme veterinare. București

Probleme zootehnice și veterinare.
București

Problèmes économiques. Paris

Problèmes de l'Europe. Paris-Rome

Problems of Economics. White Plans
New York

Problemy Ekonomiczne. Kraków

Problemy gematologii i pereliva-
niya krovi. Moskva

Problemy Kosmicheskoi Biologii,
Akademiya Nauk SSSR Moskva

Problemy Severa. USSR

Problemy Zagospodarowania Ziem
Górskich. Krakow

Problemy Zagospodarowania Ziem
Górskich. Komitet Zagospodarowania Ziem Górskich PAN. Kraków

Proceedings. Agricultural Pesticide Technical Society. London,
Ontario

Proceedings. Alaskan Science
Conference (1964). College

Proceedings of the American
Academy of Arts and Sciences.
Boston

Proceedings of the American
Association for Cancer Research.
Chicago

Proceedings. American Feed
Manufacturers Association

Proceedings of the American
Grassland Council. USA

Proceedings of the American
Philosophical Society.
Philadelphia

Proceedings of the American
Pomological Society. Washington

Proceedings. American Society for Horticultural Science. College Park. Md.

Proceedings. American Society of Sugar Beet Technologists. Salt Lake City.

Proceedings. American Veterinary Medical Association. Chicago

Proceedings of the American Wood Preservers' Association. Baltimore

Proceedings of the Animal Husbandry Wing. Board of Agriculture, India. New Delhi

Proceedings of the Annual American Association of Equine Practitioners Convention

Proceedings of the Annual Biology Colloquium. Corvallis

Proceedings of the Annual Conferences of the California Mosquito Control Association. Turlock

Proceedings of the Annual Conference of Professional Officers. Federal Ministry of Agriculture, Rhodesia and Malawi. Salisbury, Rhodesia

Proceedings of the Annual Convention of Sugar Technologists Association of India. Kanpur

Proceedings. Annual Fruit and Vegetable Short Course

Proceedings of the Annual Meeting of the American Society of Plant Physiologists at the University of Maryland. Lancaster. Pa

Proceedings of the Annual Meeting of American Wood Preservers' Association. Washington

Proceedings of the Annual Meeting of the Council on Fertilizer Application. Washington

Proceedings of the Annual Meeting of the New York State Horticultural Society

Proceedings of the Annual Meeting of the Washington State Horticultural Association, Olympia. Yakima

Proceedings of the Annual Research Conference. California Fig Institute. Fresno

Proceedings of the Association of Official Seed Analysts of North America. New Brunswick

Proceedings of the Association of Southern Agricultural Workers. Raleigh, N.C.

Proceedings of the Australian Society of Animal Production. Armidale, Melbourne

Proceedings of the Bihar Academy of Agricultural Sciences. Bangalore City, Sabow

Proceedings of the Biological Society of Washington

Proceedings of the Botanical Society of the British Isles. Arbroath, London

Proceedings British Insecticides and Fungicides Conference

Proceedings of the British Weed Control Conference. London

Proceedings. British West Indies Sugar Technologists. Barbados

Proceedings of the Burma Medical Research Society. Faculty of Medicine, University of Rangoon. Rangoon

Proceedings of the California Academy of Sciences. San Francisco

Proceedings of the California Weed Conference

Proceedings of the Canadian Centennial Wheat Symposium Saskatoon

Proceedings of the Canadian Phytopathological Society. Ottawa

Proceedings of the Canadian Society of Animal Production

Proceedings of the Central States Forest Tree Improvement Conference. USA

Proceedings of the Ceylon Association for the Advancement of Science. Colombo

Proceedings of the Chemical Society
London

Proceedings. Congres Annual.
Corporation des Ingenieurs
Forestiers, Quebec

Proceedings of the Congress of
the International Union of
Forest Research Organization.
Vienna

Proceedings of the Congress of
the South African Sugar
Technologists Association.
Durban

Proceedings. Cornell Nutrition
Conference for Feed Manufacturers

Proceedings of the Crop Science
Society of Japan. Tokyo
Proceedings of the Egyptian
Academy of Sciences. Cairo

Proceedings of the Entomological
Society of British Columbia.
Victoria, B.C. etc

Proceedings of the Entomological
Society of Manitoba. Winnipeg

Proceedings of the Entomological
Society of Ontario. Toronto

Proceedings of the Entomological
Society of Washington

Proceedings of the European
Conference for Forage Production
on Natural Grassland in Mountain
Regions. Zurich

Proceedings of the Fertilizer
Society. London

Proceedings. Florida Nutrition
Conference

Proceedings of the Florida State
Horticultural Society. Deland

Proceedings of the Forestry
Symposium. Louisiana School of
Forestry and Wildlife Management
Baton Rouge

Proceedings. General Meeting
European Grassland Federation.
Wageningen

Proceedings. Geogia Nutrition
Conference for Feed Manufacturer

Proceedings of the Grassland
Society of Southern Africa

Proceedings of the Gulf and
Caribbean Fisheries Institute.
Coral Gables

Proceedings of the Hawaiian
Academy of Science. Honolulu

Proceedings of the Hawaiian
Entomological Society. Honolulu

Proceedings of the Helminthological
Society of Washington

Proceedings of the Imperial
Academy of Japan. Tokyo

Proceedings of the Indian
Academy of Sciences. Bangalore.

Section A
Section B

Proceedings of the Indian Science
Congress

Proceedings of the Indiana
Academy of Science. Brookville

Proceedings and Informations.
Committee for Hydrological
Research TNO. The Hague

Proceedings of the International
Botanical Congress

Proceedings of the 3rd International Conference on the Peaceful
Uses of Atomic Energy. Geneva

Proceedings of the International
Conference on Scientific
Aspects of Mushroom Growing.
Philadelphia

Proceedings. International
Congress of Entomology

Proceedings. International
Congress of Genetics

Proceedings. International
Congress for Plant (Crop)
Protection

Proceedings of the International
Grassland Congress

Proceedings of the International
Horticultural Congress. Nice

Proceedings of the International
Seed Testing Association.
Vollebekk

Proceedings. International Shade
Tree Conference

Proceedings. International Society of Sugar Cane Technologists. Havana

Proceedings of the International Symposium of Soil Science. Sofia

Proceedings of the Iowa Academy of Science. Des Moines

Proceedings of the Iraqi Scientific Societies. Baghdad

Proceedings of the Japan Academy Tokyo

Proceedings of the Japanese Conference on Radioisotopes

Proceedings. K. Nederlandse akademie van wetenschappen. Amsterdam

    Series B
    Series C. Biological and
           Medical Sciences

Proceedings of the Leeds Philosophical and Literary Society. Leeds

Proceedings of the Lincoln College Farmers' Conference. Christchurch

Proceedings of the Linnean Society of New South Wales. Sydney

Proceedings of the Louisiana Academy of Sciences. Baton Rouge

Proceedings of the Malacological Society of London

Proceedings of the 2nd Malaysian Soil Conference. Kuala Lumpur

Proceedings of the Meeting Committee on Forest Tree Breeding in Canada

Proceedings. Meeting of Section. International Union of Forest Research Organizations. (IUFRO)

Proceedings of the Midwest Fertilizer Conference Washington

Proceedings. Minnesota Nutrition Conference

Proceedings of the Montana Academy of Sciences. Missoula

Proceedings of the Montpellier Symposium (UNESCO). Paris

Proceedings of the National Academy of Sciences of India. Allahabad.

    Section B. Biological Sciences

Proceedings of the National Academy of Sciences of the United States of America. Washington

Proceedings. National Conference on Clays and Clay Minerals. Washington

Proceedings of the National Institute of Sciences of India. Calcutta, New Delhi

    Ser.B. Biological sciences

Proceedings of the National Peanut Conference. USA

Proceedings of the National Symposium on Radioecology. Fort Collins.

Proceedings. New Jersey Mosquito Extermination Association. New Brunswick

Proceedings of the New South Wales Division, Australian Veterinary Association

Proceedings of the New York State Horticultural Society. Rochester

Proceedings of the New Zealand Ecological Society. Wellington

Proceedings of the 4th New Zealand Geographical Conference

Proceedings of the New Zealand Grassland Association. Palmerston North

Proceedings. New Zealand Institute of Agricultural Science. Wellington

Proceedings of the New Zealand Institute of Surveyors

Proceedings of the New Zealand Society of Animal Production. Wellington

Proceedings. New Zealand Society of Soil Science. Wellington

Proceedings. New Zealand Weed
and Pest Control Conference.
Wellington

Proceedings. North Central Weed
Control Conference. Oklahoma City

Proceedings of the North Dakota
Academy of Science. Grand Forks

Proceedings North of England Soils
Discussion Group. Durham

Proceedings. Northeastern Forest
Tree Improvement Conference

Proceedings of the Northeastern
Weed Control Conference

Proceedings of the Nutrition
Society. Cambridge. UK

Proceedings of the Nutrition
Society of Southern Africa.
Pretoria

Proceedings. Ohio State
Horticultural Society. Columbus

Proceedings of the Ohio Vegetable
Growers' Association

Proceedings of the Oklahoma
Academy of Science. Stillwater,
etc.

Proceedings of the Oregon State
Horticultural Society. Portland

Proceedings of the Pacific
Science Congress. Java, etc.

Proceedings of the Pakistan
Academy of Science

Proceedings of the Pakistan
Science Conference. Lahore

Proceedings of the Pan-Indian
Ocean Science Congress

Proceedings of the Pennsylvania
Academy of Science. Harrisburg

Proceedings of the Physical
Society. London

Proceedings of the Plant
Propagators Society.

Proceedings. Plant Science
Symposium, Campbell Soup Company
Camden, N.J.

Proceedings of the Queensland
Society of Sugar Cane Technologists. Mackay

Proceedings. R.A. Welch
Foundation

Proceedings of the Regional
Conference of the International
Potash Institute in Musten
(Switzerland). Berne

Proceedings of the Royal
Entomological Society of London.
London

Proceedings of the Royal
Institution of Great Britain.
London

Proceedings of the Royal Physical
Society of Edinburgh

Proceedings of the Royal Society.
London.

Series A.
Series B.

Proceedings of the Royal Society of
Arts and Sciences of Mauritius.
Port Louis

Proceedings of the Royal Society
of Edinburgh

Section B. Biology

Proceedings of the Royal Society
of Medicine. London

Proceedings of the Royal Society
of New Zealand. Wellington

Proceedings of the Royal Society
of Queensland. Brisbane

Proceedings of the Royal Society
of Victoria. Melbourne

Proceedings of the Ruakura
Farmers' Conference Week.
Hamilton, N.Z.

Proceedings of the Science
Association of Nigeria

Proceedings. Silvicultural
Conference, Forest Research
Institute and Colleges.
Dehra Dun

Proceedings of the Society of
American Foresters. Washington

Proceedings of the Society for
Analytical Chemistry

Proceedings of the Society for
Experimental Biology and
Medicine. New York, etc.

Proceedings. Soil and Crop Science Society of Florida. Belle Glade

Proceedings. Soil Science Society of America. Ann Arbor

Proceedings of the South African Society of Animal Production.

Proceedings of the South African Sugar Cane Technologists' Association. Durban, etc.

Proceedings of the South Dakota Academy of Sciences. Vermillion

Proceedings of the Southern Conference on Forest Tree Improvement. New Orleans

Proceedings. Southern Weed Control Conference. USA

Proceedings. Sprinkler Irrigation Association. Illinois

Proceedings of the State Horticultural Association of Pennsylvania. State College, Pa

Proceedings. Symposium on Fertility of Indian Soils. New Delhi

Proceedings. Tall Timbers Fire Ecology Conference

Proceedings of the /.../ Technical Alfalfa Conference. USA

Proceedings. Texas Nutrition Conference

Proceedings. Texas Pecan Growers' Association. College Station, Texas

Proceedings and Transactions of the Rhodesia Scientific Association. Bulawayo

Proceedings and Transactions of the Royal Society of Canada. Ottawa

Proceedings and Transactions. South London Entomological and Natural History Society. London

Proceedings. Triennnial Conference, European Association for Potato Research. Wageningen

Proceedings of the Tropical Region, American Society for Horticultural Science. Mexico

Proceedings of the United Planters Association of Southern India Annual Conference

Proceedings. United States Live Stock Sanitary Association. Chicago

Proceedings of the United States National Museum. Washington

Proceedings. University of Maryland Nutrition Conference for Feed Manufacturers.

Proceedings of the Ussher Society. Camborne

Proceedings of the Utah Academy of Sciences Arts and Letters. Salt Lake City

Proceedings. Washington State Horticultural Association. Olympia, Yakima

Proceedings of the West Virginia Academy of Science. Morganstown

Proceedings of the Western Canadian Weed Control Conference Ottawa

Proceedings. Western Forestry Conference, Western Forestry and Conservation Association. Portland, Ore.

Proceedings. Western Forest Genetics Association

Proceedings. Western Weed Control Conference

Proceedings. World Forestry Congress

Proceedings of the Zoological Society. Calcutta

Proces-verbaux des Seances de la Societe des Sciences Physiques et Naturelles de Bordeaux

Process Biochemistry. London

Processed Series of the Oklahoma Agricultural Experiment Station.

Processi verbali della Società toscana di scienze naturali in Pisa

Production Research Report.
US Department of Agriculture.
Washington

Production Yearbook. FAO Rome

Produktivnost. Beograd

Produzione animale. Italy

Professional Farm Management
Guidebook. Faculty of Agricult-
ural Economics, University of
New England, Armidale

Professional Papers. United
States Geological Survey.
Washington

Profilassi. Milano

Program Review Forest Products
Laboratory, Ottawa, Ont.

Program Review Forest Products
Laboratory, Vancouver, BC

Progrès agricole et viticole.
Montepellier

Le Progrès Scientifique. Paris

Progreso. New York

Progress. London

Progress in Biophysics and
Biophysical Chemistry. London

Progress in Biophysics and
Molecular Biology. UK/USA

Progress in Cardiovascular
Disease. New York

Progress in the Chemistry of Fats
and other Lipids. Oxford

Progress in Immunobiological
Standardization. Switzerland

Progress in Industrial
Microbiology. London

Progress in Medical Genetics.
USA/UK

Progress in Nucleic Acid Research
and Molecular Biology UK/USA

Progress Report of the
Agricultural Advisory Council

Progress Report. Branch Experiment
Station, Debre Zeit.

Progress Report. Experimental
Husbandry Farms and Experimental
Horticulture Stations. National
Agricultural Advisory Service.
London

Progress Report from Experiment
Stations. Cotton Growing
Corporation. London

Progress Report. Experiment Station
Cotton Research Corporation
Kenya

Progress Report. Experiment Station
Cotton Research Corporation
Northern State, Nigeria

Progress Report. Experiment Station
Cotton Research Corporation.
Swaziland

Progress Report. Experiment Station
Cotton Research Corporation.
Western Nigeria

Progress Report. Idaho
Agricultural Experiment Station.
Moscow, Idaho

Progress Report of New Hampshire

Progress Report. Oak Ridge
National Laboratory, Health
Physics Division

Progress Report. Pennsylvania
Agricultural Experiment Station.
State College

Progress Report of the Research
Department of the Ministry of
Agriculture and Fisheries.
Barbados

Progress Report on Research and
Technical Work, Ministry of
Agriculture, Northern Ireland.
Belfast

Progress Report Series of the
Auburn University Agricultural
Experiment Station.

Progress Report of the Texas
Agricultural Experiment Station.
College Station.

Progress Report (or Summary
Report) Tree Nursery. Indian
Head, Sask.

Progress Report. Tropical
Pesticides Research Institute.

Progress Report. United States
Atomic Energy Commission.
Health Physics Division.
Washington

Progress Report. University of
Kentucky Agricultural Experiment
Station. Lexington

Progress in Soil Biology

Progress in Soil Zoology. London

Progressive Agriculture in
Arizona. Tucson

Progressive Fish Culturist. Fish
and Wildlife Service.
Washington

Progresso agricolo. Bologna

Progresso medico. Napoli
Progresso veterinario. Torino

Projet. Paris

Promotions. Paris

Protoplasma. Leipzig, etc

Průmysl potravin. Praha

Przegląd epidemiologiczny.
Warszawa

Przegląd geofizyczny. Warszawa

Przegląd geograficzny. Warszawa

Przegląd hodowlany. Warszawa

Przegląd lekarski. Krakow,Warszawa

Przegląd mleczarski. Warszawa

Przegląd zoologiczny. Wrocław

Przemysł chemiczny. Warszawa

Przemysł drzewny. Warszawa

Przemysł spozywczy. Warszawa

Psyche, a journal of Entomology.
Cambridge, Mass.

Psychiatrie, Neurologie and
Medizinische Psychologie.
Berlin, Leipzig

Psychosomatic Medicine.
Washington, Philadelphia, etc.

Ptitsevodstvo. Moskva

Pubblicazioni del Centro di
sperimentazione agricola e
forestale. Roma

Pubblicazioni. Ente nazionale per
la cellulosa e per la carta.
Roma

Pubblicazioni dell'Istituto di
patologia vegetale, Università
di Milano. Milano

Pubblicazioni stazione sperimentale
di selvicoltura. Firenze

Public Health. Johannesburg

Public Health. London

Public Health Papers. WHO Geneva

Public Health Reports. Washington

Public Roads. US Department of
Agriculture. Washington

Publicación, Dirección General del
Inventario Nacional Forestal,
Coyoacán, Mexico

Publicacion. Instituto Nacional de
Tecnologia Agropecuria (INTA)
Buenos Aires

Publicacion del Instituto de
Selva. Lima

Publicacion Servicio de Plagas
Forestales, Madrid

Publicaciones. Escuela de
veterinaria, Universidad de
Buenos Aires

Publicaciones especiales.
Instituto Nacional de investigaciones Forestales Secretaria
de Agricultura y Ganaderia.
Mexico

Publicaciones del Instituto de
biologia aplicada. Barcelona

Publicaciones. Instituto mexicana
de recursos renovables. Mexico
(numbered series)

Publicaciones. Instituto de
Suelos y Agrotecnia. Buenos
Aires

Publicaciones miscelaneas.
Estacion experimental agricola.
Tucuman

Publicaciones del Museo de
historia natural "Javier Prado"
Lima

Publicaciones Ocasionales del
Museo de Ciencias Naturales,
Caracas.

Zoologia

Publicaciones técnicas. Estación
experimental de Pergamino.
Pergamino

Publicações avulsas do Instituto
Centro de Pesquisas, Aggeu
Magalhães. Recife

Publicações culturais da
Companhia de diamantes de
Angola. Museu do Dundo. Lisboa

Publicações. Direcção geral dos
serviços florestais e aquicolas.
Lisboa

Publicações. Instituto de
micologia, Universidade do
Recife. Recife.

Publicações, Serviços de Agricultura, Serviços de Veterinária,
Lourenço Marques

Publicatiës van het Natuurhistorisch genootschap in Limburg.
Maastricht

Publication. Agricultural Adjustment Unit, University of
Newcastle upon Tyne.

Publications of the American
Association for the Advancement
of Science. New York

Publications. American University
of Beirut. Faculty of
Agricultural Sciences. Beirut

Publications. Australian Society
of Soil Science. Canberra

Publication. Board of Economic
Inquiry Punjab. Lahore

Publications. Canada Department
of Agriculture. Ottawa

Publications. Canada Department
of Agriculture, Research Branch.
Ottawa

Publications. Canada Department
of Forestry. Ottawa

Publications. Centre d'étude
pour l'utilisation des sciures
de bois. Louvain

Publications. Centre technique
forestier tropical. Nogent-
sur-Marne

Publication. Cooperative Extension
Service, College of Agriculture,
Washington State University,
Pullman, Washington

Publication. Cornell University
Agricultural Experiment Station.
Ithaca. N.Y.

Publication. Department of
Agricultural Economics, New
York State College of Agricul-
ture, Ithaca, N.Y.

Publication. Department of
Agricultural Economics and
Rural Sociology, South Carolina
Experiment Station, Clemson
University, Clemson, SC.

Publication. Department of
Agricultural Economics, University of Illinois, Urbana, Ill.

Publications. Department of
Agriculture, Canada. Ottawa

Publication. Economic Research
Service, US Department of
Agriculture, Washington

Publication. Economic Statistical
Analysis Division, Economic
Research Service, US Department
of Agriculture, Washington

Publications. European and
Mediterranean Plant Protection
Organization. Paris

Publications of the Faculty of
Agricultural Sciences, American
University of Beirut. Beirut

Publication. FAO/ECE Joint
Committee on Forest Working
Techniques and Training of
Forest Workers

Publication. Foreign Regional
Analysis Division, Economic
Research Service, US Department
of Agriculture, Washington

Publications. Forest Research
Branch. Ottawa

Publication. Hawaii Agricultural
Experiment Station, Honolulu

Publications de l'Institut
National pour l'Etude Agronomique du Congo. Bruxelles

Serie Scientifique
Serie Technique

Publications de l'Institut
technique français de la
betterave industrielle. Paris

Publications of the Institute of
Marine Science, University of
Texas. Austin

Publications. International
Association of Scientific
Hydrology. General Assembly of
Berkeley. Gentbrugge

Publications. International
Association of Scientific
Hydrology. Symposium of Bari.
Gentbrugge

Publications. International
Institute for Land Reclamation
and Improvement. Wageningen

Publication of the Land
Capability Survey of Trinidad
and Tobago

Publications. Manitoba Department
of Agriculture. Winnipeg

Publications of the Maria Moors
Cabot Foundation for Botanical
Research. Petersham, Mass

Publications of the Marketing
Research Institute of Pellervo
Society, Helsinki

Publications. National Academy of
Sciences, National Research
Council. Washington

Publication. National Council for
Research and Development. Water
Authority, Ministry of Agricul-
ture. Weizmann Science Press of
Israel

Publication. New York State
Agricultural Experiment Station
Cornell University. Ithaca, NY

Publication. New York State
College of Agriculture, Ithaca.

Publication. New Zealand Meat and
Wool Board's Economic Service,
Wellington

Publication. Nobiyuku Nogyo
Noseichosoiikai. Tokyo

Publications. Ontario Department
of Agriculture. Toronto

Publications and Patents.
Eastern Utilization Research and
Development Division.
Philadelphia

Publications du Service de la
carte géologique de
Luxembourg

Publications of the Seto Marine
Biological Laboratory.
Sirahama

Publications. Soil Bureau
Department of Scientific and
Industrial Research. Wellington

Publication. South Carolina
Agricultural Experiment Station.
Clemson, S.C.

Publication. Supply Division
New Zealand Dairy Board.
Wellington

Publications. Station federales
d'essais agricoles. Lausanne

Publications techniques de
l'Institut belge pour
l'amelioration de la
betterave. Bruxelles

Publication. Texas Agricultural
Experiment Station, College
Station. Texas

Publications. University of
Ankara Faculty of Agriculture.
Ankara

Publications de vulgarisation de
l'Institut belge pour l'amelioration de la betterave. Tirlemont

Publication. World Food Problem
Food and Agricultural Organization. Rome

Publicazione Instituto di
Economia e Politica Agraria del'
Universita degli Studi di Parma.
Parma

Publikatie. Instituut voor Land-
bouwtechniek en Rationalisatie.
Wageningen

Publikatie. Landbouw-Economische
Instituut. 's Gravenhage

Publikatie. Proefstation voor de
Akkeren Weidebouw. Wageningen

Publikatie van het Proefstation
voor de Groeten- en Fruitteelt
onder Glas te Naaldwijk

Publikation. Statens Offentliga
Utredningar. Stockholm

Pulp and Paper Magazine of
Canada. Montreal

Punjab Horticultural Journal

Pure and Applied Chemistry.
London

Purpan. Toulouse

Puutarha. Tamperessa, etc

Pyrethrum Post. London, Nakuru,
Kenya

Q

Quaderni. Laboratorio
crittogamico, Istituto botanico
della Universita, Pavia

Quaderni della nutrizione.
Istituto di fisiologia generale,
Citta Universitaria. Roma

Quaestiones Entomologicae,
Edmonton, Canada

Quail Quarterly. USA

Qualitas plantarum et materiae
vegetabiles. Den Haag

Quality Publication, Malting
Barley Improvement Association.
Milwaukee

Quarterly Bulletin of Economics
and Statistics. Bombay

Quarterly Bulletin. International
Association of Agricultural
Librarians and Documentalists.
Harpenden, Wageningen

Quarterly Bulletin. Michigan
State University Agricultural
Experiment Station. East Lansing

Quarterly Bulletin of North-
western University Medical
School. Evanston

Quarterly Circular. Ceylon Rubber
Research Scheme. Peradeniya

Quarterly Journal of Experimental
Physiology. Edinburgh

Quarterly Journal of Experimental
Physiology and cognate medical
sciences. London

Quarterly Journal. Florida
Academy of Sciences. Gainesville
Tallahassee

Quarterly Journal of Forestry.

Quarterly Journal of Medicine.
Oxford

Quarterly Journal of the Royal
Meteorological Society. London

Quarterly Journal. Rubber
Research Institute of Ceylon.
Colombo

Quarterly Journal of Studies on
Alcohol. New Haven, Conn.

Quarterly Journal of the Taiwan
Museum. Taipei

Quarterly Progress Report. West
African Cocoa Research Institute
Tafo

Quarterly Progress Report.
Nigerian Institute for Oil Palm
Research. Benin City

Quarterly Report. FAO Plant
Protection Committee for the
South East Asia and Pacific
Region. Bangkok

Quarterly Report. University of
the West Indies School of
Agriculture. Trinidad

Quarterly Research Bulletin.
Federal Department of Agricultural Research. Ibadan

Quarterly Review of Agricultural
Economics. Canberra

Quarterly Review. Banco Nazionale
di Lavore. Roma

Quarterly Review of Biology.
Baltimore

Quarterly Review of Biophysics.
Cambridge

Quarterly Review. Chemical
Society. London

Quarterly Review. National
Westminster Bank, London

Quarterly Review of Scientific
Publications, Polish Academy
of Sciences. Warsaw

Quarterly Review. Skandinaviska
Banken. Stockholm

Quarterly Review. Veterinary
Institute for Tropical and High
Altitude Research. Peru

Quarterly. Serving farm ranch and
home. University of Nebraska,
College of Agriculture and Home
Economics.

Quebec laitier et alimentaire.
Quebec

Queensland Agricultural Journal.
Brisbane

Queensland Journal of
Agricultural and Animal
Sciences. Brisbane

Quellen und Studien des Instituts
für Genossenschaftswesen der
Universität Münster. Münster

R

R.A.S.E. Review. UK

R.R.I.C. Bulletin. N.S. Dartonfield

Raboty Tyan'shanskoi fiziko-
geograficheskoi stantsii.
Akademiya Nauk Kirgizskoi SSR.
Frunze

Race Hygiene. Tokyo

Radiation Botany. Oxford

Radiation Research. New York

Radiobiologiya

Radiography. London

Radioisotopes in Soil Plant
Nutrition Studies. Proceedings
of a Symposium. Bombay

Radio-Isotopes. Tokyo

Radiokhimiya. Moskva

Radiologia clinica. Berlin, Bern

Radiologia medica. Torino

Radiological Health Data and
Reports. Washington

Radiology. St. Paul, etc.

Radovi, Institut za Šumarska
Istrazivanja, Šumarskog
Fakulteta, Sveučilišta u
Zagrebu, Zagreb

Radovi Poljoprivrednog Fakulteta
Univerziteta u Sarajevu.
Yugoslavia

Radovi Poljoprivredno šumarskog
fakulteta Univerziteta u
Sarajevu. Sarajevo

Radovi. Sumarski Fakultet i
Institut za Sumarstvo u
Sarajevu, Sarajevo

Range Improvement Studies.
California Division of
Forestry. Sacramento

Rapport d'Activité. Centre de
Recherches Agronomiques de
l'Etat. Gembloux. Belgium

Rapport d'activité. Laboratoire
forestier de l'etat. Gembloux

Rapport d'activite. Station
d'amelioration des plantes
maraichers. Montfavet

Rapport Annuel, Centre Technique
Forestier Tropical, Nogent-sur-
Marne

Rapport annuel. Institut pour la
recherche scientifique en
Afrique centrale. Bruxelles

Rapport. Centre technique
interprofessionnel des
Oléagineux Metropolitains.
France

Rapport. Commissariat à l'énergie
atomique (France). Gif-sur-
Yvette

Rapport sur le fonctionnement de
l'institut Pasteur du Maroc.
Casablanca

Rapport Général. Fifth World
Fertilizer Congress. CITA.
Zurich

Rapport. Institut de Recherches
Agronomiques Tropicales et des
Cultures Vivrières. Saint-Denis,
Île de la Réunion

Rapport. Nederlands Instituut
voor Zuivelonderzoek. Ede

Rapport. Société d'encouragement
de la culture des orges de
brasserie en France. Secobra

Rapport de la Station d'amelioration des plantes de Dijon

Rapport de la Station d'amelioration des plantes de Montpellier

Rapport de la Station d'amelioration des plantes de Rennes

Rapport de la Station Centrale de Genetique et d'amelioration des plantes

Rapporto della stazione sperimentale di praticoltura di Lodi

Rapporter och Uppsatser Institutionen for Skoglig Matematisk Statistik.

Rapporter och Uppsatser Institutionen for Skogsekologi.

Rapporter och Uppsatser, Institutionen för Skogsföryngring, Skogshögskolan, Stockholm

Rapporter och Uppsatser, Institutionen för Skogsgenetik, Skogshögskolan, Stockholm

Rapporter och Uppsatser Institutionen for Skogsproduktion

Rapporter och Uppsatser Institutionen for Skogstaxering

Rapporter och Uppsatser Institutionen for Skogsteknik

Rapporter. Institutionen for Virkeslara. Skogshogskolan

Rapports speciaux. World Fertilizer Congress. CITA. Zurich

Rassegna Chimica. Rome

Rassegna di clinica, terapin e scienze affini. Roma

Rassegna medica sarda. Cagliari

Rassegna di medicina industriale e di igiene del lavoro. Torino, Roma

Rassegna pugliese di tecnica vinicola e agraria. Bari

Rastenievadni nauki. Akademiya na selskostopanskite nauki. Sofia

Rastitel' nye rosursy. Moskva

Rastitelna zashtita. Sofiya

Ratgeber sozialistischer Landwirtschaft. Magdeburg

Raumforschung und Raumordnuug. Bad Godesberg, Köln

Realta del mezzogiorno. Roma

Recent Progress in Hormone Research. New York

Recent Progress in Microbiology. Symposia VIII International Contress for Microbiology, Montreal. Toronto

Recherches agronomiques. Quebec

Recherches economiques de Louvain. Louvain

Recherche Sociale. Paris

Recherches Veterinaires. France

Reclamation Era. Washington

Record of Agricultural Research, Northern Ireland. Belfast

Record of the Annual Convention of the British Wood Preservation Association. London

Record of Investigations. Department of Agriculture, Uganda. Entebbe

Record of Researches in the Faculty of Agriculture, University of Tokyo. Tokyo

Record. Scottish Plant Breeding Station. Roslin

Records of the Australian Museum

Records of the Canterbury Museum Christchurch

Records of the Dominion Museum Wellington

Records of the Indian Museum Calcutta, Delhi

Records of the South Australian Museum

Recueil de médecine vétérinaire

Recueil des travaux chimiques des Pays-Bas et de la Belgique. Leyde, Amsterdam, etc.

Recueil des travaux du Laboratoire de biologie végétale, Faculté des sciences de Bordeaux. Bordeaux

Recueil des travaux des Laboratoires de botanique, geologie et zoologie de la Faculte des Sciences de l'Universite de Montpellier

Serie Zoologique

Redia. Giornale de Entomologia Florence

Redogörelse, Forskningsstiftelsen
Skogsarbeten. Stockholm

Referativnyi Zhurnal. Moskva

Entomologiya
Biologiya
Rastenievodstvo. Otdel'nyi
Vypusk
Zhivotnovodstvo. Waterinariya

Reflets et perspectives de la vie
economique. Bruxelles

Refuah veterinarith. Jerusalem,
Tel Aviv

Regional Research Centre of the
British Caribbean at the
Imperial College of Tropical
Agriculture. Trinidad

Regional Science Association
Papers. Philadelphia

Regnum vegetabile. Utrecht

Reichenbachia. Dresden

Rendiconti dell'accademia
nazionale die Lincei: classe di
scienze fische, matematiche e
naturali.

Rendiconti dell'Istituto lombardo
di scienze e lettere. Milano

B. Scienze biologiche e
mediche

Rendiconti della Società
mineralogica italiana. Pavia

Report and Accounts. Tanzania
Agricultural Corporation

Report on the Activities of the
Danish Atomic Energy
Commission. Copenhagen

Report on Advisory Services and
Howard Davis Farm, Jersey.
St. Helier

Report of the Agricultural Branch,
Northern Territory Administration
Australia

Report of the Agricultural
Department, Antigua. St. John

Report of the Agricultural
Department, British Virgin
Islands. Tortola

Report on the Agricultural
Department, University College,
Dublin

Report of the Agricultural
Division. Eastern Region of
Nigeria

Report of the Agricultural
Division. Ministry of
Agriculture, Forests and
Wildlife. Tanzania

Report. Agricultural Experiment
Stations. Institute of Food and
Agricultural Sciences. University
of Florida. Gainesville

Report of the Agricultural and
Horticultural Research Station,
University of Bristol. Long
Ashton

Report. Agricultural Research
Council of Central Africa.
Salisbury

Report. Agricultural Research
Division. Ministry of
Agriculture. Republic of the
Sudan

Report. Agricultural Research
Council. London

Report. Agricultural Research
Council. Letcombe Laboratory.
Wantage

Report. Agricultural Research
Institute. Cyprus. Nicosia

Report. Agricultural Research
Institute of Northern Ireland.
Hillsborough, Co.Down

Report. Agricultural Research
Station, Kpong , University of
Ghana

Report. Alberta Soil Survey.
University of Alberta.
Edmonton

Report. Alfalfa Improvement
Conference. USA

Report of the Ankara Agricultural
Research Institute. Ankara

Report. Animal Breeding Research
Organization. Edinburgh

Report of the Animal Health
Research Centre, Entebbe

Report on Animal Health Services
in Great Britain. Ministry of
Agriculture, Fisheries and Food.
London

Report of the Annual Date Growers'
Institute.

Report of the Annual Meeting.
Massachusetts Fruit Growers'
Association. Amherst

Report. Arizona Agricultural
Experiment Station. Tucson

Report. Arizona Commission of
Agriculture and Horticulture.
Phoenix

Report of the Arkansas Agricultural
Experiment Station. Fayetteville

Report. Arthur Richwood
Experimental Husbandry Farm

Report. Ayub Agricultural
Research Institute.

Report. Banana Board Research
Department, Jamaica. Kingston

Report on Barley Committee,
European Brewery Convention.
Rotterdam

Report of the Bilharzia Snail
Control Section, Ministry of
Public Health, Egypt. Cairo

Report of the Board of Regents of
the Smithsonian Institution.
Washington

Report of the Bose Research
Institute. Calcutta

Report. Botanic Gardens Department.
Singapore

Report. Boxworth Experimental
Husbandry Farm.

Report. Bridgets Experimental
Husbandry Farm

Report. British West Indies
Sugar Association

Report. Bureau of Sugar Experiment
Stations, Queensland. Brisbane

Report on Cacao Research.
Regional Research Centre
Imperial College of Tropical
Agriculture, University of
West Indies. St. Augustine

Report. Cawthron Institute of
Scientific Research. Nelson, NZ

Report. Central Rice Research
Institute. Cuttack

Report of the Central Tobacco
Research Institute. Rajahmundry

Report. Chief Veterinary Officer
Cyprus. Nicosia

Report. Cocoa Conference. London

Report. Cocoa Research Institute
Ghana Academy of Sciences. Tafo

Report. Cocoa Research Institute
of Nigeria. Ibadan

Report. Coconut Industry Board.
Jamaica

Report of the Coffee Board.
Bangalore

Report. Coffee Research and
Experimental Station, Lyamungu,
Moshi, Tanzania. Dar-es-Salaam

Report. Coffee Research
Foundation Station, Kenya, Ruiru

Report of the Colorado Agricultural
Experiment Station; Fort Collins

Report. Colonial Development
Corporation. London

Report. Colonial Pesticides
Research Committee. London

Report. Committee for Colonial
Agriculture, Animal Health and
Forestry Research. London

Report. Commonwealth Agricultural
Bureaux. London, Farnham Royal

Report. Commonwealth Forestry
Institute. University of Oxford

Report. Commonwealth Institute of
Biological Control. Trinidad

Report of the Commonwealth
Scientific and Industrial
Research Organization,
Australia. Canberra, Melbourne

Report of the 6th Conference of
the International Association
on Quaternary Research. Warsaw

Report. Cooperative Industry, NC
State Harwood Research Program.
North Carolina State University
School of Forestry. Releigh

Report. Cornell University College
of Agriculture and Experiment
Station. Ithaca

Report. Cotton Research
Corporation. London

Report. Council for Scientific and
Industrial Research, Union of
South Africa. Pretoria

Report. Crop Husbandry Department,
North of Scotland College of
Agriculture.

Report on Crop Research in
Lesotho. Maseru

Report. Cyprus Agricultural
Research Institute. Nicosia

Report of the Date Growers'
Institute. Indio

Report. Deir Alla Research
Station. Jordan

Report. Department of Agricultural
Economics, North Dakota
Agricultural Experiment Station.
Fargo. N.Dakota

Report. Department of Agricultural
Economics, University of
Exeter, Exeter

Report. Department of Agricultural
Economics, University of
Newcastle upon Tyne.

Report. Department of Agricultural
Economics, University of
Nottingham, Sutton Bonington

Report of the Department of
Agricultural Marketing,
University ofNewcastle, Newcastle

Report. Department of Agricultural
Research, Federation of
Nigeria. Lagos

Report. Department of Agricultural
Technical Services, Pretoria

Report of the Department of
Agriculture, Alberta. Edmonton

Report. Department of Agriculture,
Bermuda. Hamilton

Report. Department of Agriculture,
Botswana. Vryburg, etc

Report. Department of Agriculture,
British Columbia, Victoria

Report. Department of Agriculture,
British Honduras. Belize

Report. Department of Agriculture,
Canada

Report. Department of Agriculture,
Ceylon. Colombo

Report. Department of Agriculture
and Conservation, Hawaii.
Honolulu

Report. Department of Agriculture,
Cyprus. Nicosia

Report of the Department of
Agriculture, Federation of
Malaysia. Kuala Lumpur

Report. Department of Agriculture,
Fiji. Suva

Report. Department of Agriculture,
Forest and Fisheries, Western
Samoa. Apia

Report. Department of Agriculture
and Forestry, Dominica, Roseau

Report. Department of Agriculture
and Forestry, St. Helena

Report. Department of Agriculture
and Irrigation. South Arabia

Report on the Department of
Agriculture Jamaica. Kingston

Report. Department of Agriculture,
Kenya. Nairobi

Report. Department of Agriculture,
Lesotho. Bloemfontein

Report. Department of Agriculture,
Malawi. Zomba

Report. Department of Agriculture,
Malta. Valetta

Report of the Department of
Agriculture. Mauritius. Port
Louis

Report. Department of Agriculture,
New Brunswick. Fredericton

Report. Department of Agriculture,
New South Wales. Sydney

Report. Department of Agriculture,
New Zealand. Wellington

Report. Department of Agriculture,
Ontario

Report. Department of Agriculture,
Zambia. Lusaka

Report. Department of Agriculture,
Sabah

Report. Department of Agriculture,
Sarawak. Kuching

Report. Department of Agriculture,
Seychelles Islands. Victoria,
Mahe

Report of the Department of
Agriculture. Sierra Leone.
Freetown

Report of the Department of
Agriculture, Solomon Islands.
Honiara

Report of the Department of
Agriculture, South Australia.
Adelaide

Report of the Department of
Agriculture and Stock,
Queensland. Brisbane

Report. Department of Agriculture,
Stock and Fisheries, Territory
of Papua and New Guinea

Report. Department of Agriculture,
Swaziland

Report. Department of Agriculture,
Tanzania. London and Dar-es-
Salaam

Report. Department of Agriculture,
Tasmania. Hobart

Report of the Department of
Agriculture, Trinidad and Tobago
Port of Spain

Report of the Department of
Agriculture, Uganda. Entebbe

Report of the Department of
Agriculture, Western Australia.
Perth

Report of the Department of
Agriculture, Zanzibar

Report. Department of Economics
(Agricultural Economics).
University of Exeter, Exeter

Report. Department of Hop
Research, Wye College. Wye

Report. Department of Lands,
South Australia

Report on the Department of
Medical Services of the Northern
Region of Nigeria. Kaduna

Report of the Department of
Primary Industries, Queensland.
Brisbane

Report of the Department of
Science and Agriculture,
Barbados. Bridgetown

Report. Department of Scientific
and Industrial Research,
New Zealand. Wellington

Report of the Department of
Veterinary Services, Kenya,
Nairobi

Report. Department of Veterinary
Services and Animal Industry,
Malawi. Zomba

Report. Department of Veterinary
Services and Animal Industry,
Tanzania

Report. Department of Veterinary
Services and Animal Industry,
Uganda, Entebbe

Report on the Department of
Veterinary Services of the
Northern Region of Nigeria.
Kaduna

Report of the Desert Locust
Organization for Eastern Africa.
Nairobi

Report. Director of Agriculture
and Fisheries, Bermuda.
Hamilton

Report of the Director of
Agriculture, Cyprus. Nicosia

Report of the Director, Anti-
Locust Research Centre. London

Report of the Director of the
Department of Plant Biology,
Carnegie Institution. Washington

Report. Director of Forest
Products Research. London

Report of the Director, International Red Locust Control
Service, Abercorn. Zambia

Report. Director of Medical
Services, Medical Department,
British Guiana. Georgetown

Report of the Directorate of Plant
Protection, Quarantine and
Storage. India

Report. Directorate of Overseas
(Geodetic and Topographical)
Surveys. UK

Report of the Ditton and Covent
Garden Laboratories. UK

Report of the Division of
Agriculture, Federation of
Malaysia. Kuala Lumpur

Report of the Division of
Agriculture, Sierra Leone,
Freetown

Report. Division of Forest
Products, Melbourne, Australia

Report. Division of Laboratories
and Research, New York State
Department of Health. Albany

Report. Division of Plant Industry
CSIRO, Australia. Canberra

Report. Division of Tropical
Pastures, CSIRO, Australia.
Melbourne

Report. Drayton Experimental
Husbandry Farm

Report. East African Agriculture
and Forestry Research
Organisation. Nairobi

Report. East African Institute of
Malaria and Vector-borne
Diseases. Mwanza, Dar-es-Salaam

Report. East African Institute for
Medical Research. Nairobi

Report. East African
Trypanosomiasis Research. Tororo
Uganda

Report. East African Veterinary
Research Organization. Nairobi

Report. East African Virus
Research Institute. Nairobi

Report of the East Malling
Research Station. East Malling

Report. Economic Research
Service, US Department of
Agriculture, Washington

Report. Efford Experimental
Horticulture Station.

Report. Eley Game Advisory
Station, Fordingbridge, Hants

Report of the Experiment Station
Committee, Hawaiian Sugar
Planters' Association. Honolulu

Report. Experiment Station of the
South African Sugar Association
Mount Edgecombe

Report. Experiment Stations. University of Georgia College of
Agriculture. Athens

Report of the FAO/IAEA Technical
Meeting. Brunswick-Völkenrode
(1963) Pergamon Press

Report of the Faculty of
Agriculture, Okayama University.
Okayama

Report. Fairfield Experimental
Horticulture Station. Preston

Report of the Federal Department
of Veterinary Research,
Federation of Nigeria. Lagos

Report. Filariasis Research
Unit, East Africa

Report of the Florida Agricultural
Experiment Station. Gainesville

Report of the Food Research
Institute Aichi Prefecture.
Aichi, Japan

Report. Food Research Institute.
Tokyo

Report. Foreign Region Analysis
Service, US Department of
Agriculture, Washington

Report on Forest Administration
Federation of Malaysia. Kuala
Lumpur

Report. Forest Department,
British Honduras. Belize

Report. Forest Entomology and
Pathology Branch, Canada
Department of Forestry. Ottawa

Report of the Forest Insect and
Disease Survey. Forest
Entomology and Pathology Branch.
Department of Forestry, Canada.
Ottawa

Report of the Forest Insect
Survey in the Province of
Quebec

Report. Forest Products Laboratory
Madison, Wis.

Report. Forest Products Research
Forest Research Laboratory.
Oregon

Report. Forest Products Research
Institute College. Laguna

Report on Forest Research.
Forestry Commission. London

Report. Forest Research Glendon
Hall. Faculty of Forestry.
Toronto University

Report. Forest Research Institute,
New Zealand Forest Service.
Whakarewarewa

Report. Forest Resources
Reconnaissance Survey of
Malaya. Kepong

Report. Forest Soils of Japan
Government Forest Experiment
Station. Meguro

Report of the Forestry Commission
London

Report of the Forestry Commission,
New South Wales. Sydney

Report. Forestry and Timber
Bureau. Canberra

Report. Forests Commission,
Victoria. Melbourne

Report of the Forests Department
Western Australia. Perth

Report of the Gatooma Research
Station. Rhodesia

Report of the Georgia Agricultural
Experiment Station. Athens

Report. Georgia Forest Research
Council. Macon

Report of the Gezira Research
Station. Sudan

Report of the Glasshouse Crops
Research Institute.
Littlehampton

Report. Gleadthorpe Experimental
Husbandry Farm

Report. Gorgas Memorial Laboratory
Institute of Tropical Medicine,
Panama, Washington

Report. Grasslands Agricultural
Research Station, Marandellas,
Rhodesia

Report of the Grassland Research
Institute. Hurley

Report. Hannah Dairy Research
Institute. Glasgow

Report of the Hawaii Agricultural
Experiment Station. Washington,
Honolulu

Report. Hawaiian Sugar Planters'
Association Experiment Station.
Honolulu

Report. Hawaiian Sugar
Technologists. Honolulu

Report of the Hebrew University of
Jerusalem

Report of the Hellenic Agricultural
Research Station. Larissa

Report. Henderson Research Station,
Mazoe. Rhodesia

Report of the Herbicide Research
Institute, University of West
Indies. Trinidad

Report. High Mowthorpe Experimental
Husbandry Farm

Report. Hill Farming Research
Organization. Edinburgh

Report of the Hokkaido Branch,
Government Forest Experiment
Station. Sapporo

Report. Hokkaido Forest Products
Research Institute. Hokkaido

Report of the Hokkaido National
Agricultural Experiment Station.
Sapporo

Report. The Horticultural Centre,
Loughgall. Ministry of
Agriculture, Northern Ireland.
Loughgall

Report of the Illinois Agricultural
Experiment Station. Urbana

Report. Imperial Ethiopian
Government Institute of
Agricultural Research. Abbis Ababa

Report. Indian Central Institute
for Communicable Diseases. Delhi

Report. Indian Coffee Board.
Bangalore

Report. Indian Council of
Agricultural Research.
Calcutta, New Delhi

Report of the Indian Institute for Sugarcane Research. Lucknow

Report of the Indian Veterinary Research Institute, Muktesar and Izatnager.

Report of the Insect Vector Control Division, Health Department, Trinidad Government. Port of Spain

Report. Institute for Agricultural Research and Special Services. Ahmadu Bello University, Northern Nigeria. Samaru

Report of the Institute of Agricultural Research, Tohoku University. Sendai

Series D Agriculture

Report. Institute for Land and Water Management Research. Wageningen

Report of the Institute for Medical Research, Federation of Malaysia. Kuala Lumpur

Report of the Institute for Science of Labour. Tokyo

Report. Inter-American Institute of Agricultural Sciences of the Organization of American States Tropical Center. Washington

Report to the International Conference of Soil Scientists. New Zealand 1962

Report. International Crop Improvement Association. St. Paul

Report. International Institute for Land Reclamation and Improvement. Wageningen

Report. International Labour Office. Geneva

Report. International Rice Research Institute. Manila

Report. Iowa Agricultural and Home Economics Experiment Station. Ames

Report. Jamaica Agricultural Society. Kingston

Report. John Innes Institute. Norwich

Report of the Jute Agricultural Research Institute. Indian Central Jute Committee. Calcutta

Report of the Kansas Agricultural Experiment Station, Manhattan. Topeka

Report of the Kentucky Agricultural Experiment Station of the University of Kentucky

Report. Kirton Experimental Horticulture Station.

Report of the Kyushu University Forests, Faculty of Agriculture. Fukuoka

Report. Liscombe Experimental Husbandry Farm

Report of the Livestock Research Division, Ministry of Agriculture Tanzania

Report of the London School of Hygiene and Tropical Medicine. London

Report. Long Ashton Agricultural and Horticultural Research Station.

Report of the Louisiana Rice Experiment Station. USA

Report. Low Temperature Research Station. Cambridge

Report. Luddington Experimental Horticulture Station. Stratford on Avon. London

Report. Macaulay Institute for Soil Research. Aberdeen

Report. Makerere University College. Uganda

Report of the Malaria Advisory Board, States of Malaya. Kuala Lumpur

Report of the Maryland Agricultural Experiment Station. College Park

Report of the Matopos Research Station. Rhodesia

Report. Mauritius Sugar Industry Research Institute. Port Louis, Reduit

Report of the Mechanical Development Committee. Forestry Commission, London

Report of the Medical Department Cyprus. Nicosia

Report on Medical and Health Services, Sierra Leone

Report of the Medical Services. Ministry of Health, Sudan. Khartoum

Report of the Michigan State Horticultural Society. Lansing

Report. Milk Recording Service Department. UK

Report of the Minister of Agriculture for Canada. Ottawa

Report of the Minister for Agriculture, Eire. Dublin

Report of the Minister for Lands and Forestry, Department of Lands. Dublin

Report of the Ministry of Agriculture and Natural Resources, British Guyana. Georgetown

Report of the Ministry of Agriculture. Ghana

Report of the Ministry of Agriculture. Republic of Zambia. Lusaka

Report. Ministry of Agriculture and Lands, Jamaica. Kingston

Report of the Ministry of Agriculture of the Northern Region of Nigeria

Report. Ministry of Agriculture, Food and Cooperatives. Tanzania

Report. Ministry of Health, Northern Region, Nigeria. Kaduna

Report. Ministry of Health, Rhodesia. Salisbury

Report. Ministry of Health, Uganda Entebbe

Report. Mississippi Agricultural Experiment Station. Agricultural College

Report of the National Agricultural Advisory Service Experimental Husbandry Farms and Experimental Horticulture Stations.

Report. National Agricultural Research Station, Kitale. Kenya

Report. National Board for Prices and Incomes. London

Report. National Environmental Research Council. UK

Report of the National Institute of Agricultural Botany. Cambridge

Report. National Institute for Agricultural Engineering. Silsoe

Report. National Institute of Agricultural Engineering and Scottish Machinery Testing Station. Silsoe

Report of the National Institute for Research in Dairying, Reading University. Shinfield Reading

Report of the National Research Council, Canada. Ottawa

Report. National Vegetable Research Station. Wellesbourne

Report. Nebraska Agricultural Experiment Station. Lincoln

Report of the New Mexico Agricultural Experiment Station. College of Agriculture and Mechanic Arts. College Station

Report of the New York State Agricultural Experiment Station. Geneva, N.Y.

Report of the New York State College of Agriculture at Cornell University Agricultural Experiment Station. Albany

Report. New Zealand Forest Service, Wellington

Report of the Nigerian Institute for Oil Palm Research, nr. Benin City

Report. Nigerian Institute for Trypanosomiasis Research. Kaduna

Report. Nigerian Stored Products Research Institute. Lagos

Report of the Noda Institute for Scientific Research. Noda-shi

Report. Norfolk Agricultural Station. Norwich, etc.

Report of the North Carolina State College School of Agriculture

Report. North Carolina State Industry Co-operative Tree Improvement Program. N.C. State University. School of Forestry. Raleigh

Report of the Northeast Louisiana Agricultural Experiment Station

Report of the Northern Counties Animal Diseases Research Fund. Newcastle

Report. Northern Nigerian Development Corporation. Kaduna

Report. Northern Nigerian Institute of Agricultural Research. Samaru

Report of the Northern Nut Growers' Association

Report of the Nova Scotia Department of Agriculture. Halifax

Report. Nova Scotia Fruit Growers Association. Kentville

Report of the Ohio Agricultural Experiment Station. Wooster

Report. Oklahoma Agricultural Experiment Station. Stillwater

Report. Ontario Horticultural Experiment Stations and Products Laboratory

Report. Oregon State University Forest Research Laboratory. Corvallis

Report of the Orient Hospital. Beirut

Report. Pacific Northwest Forest Experiment Station and Range Experiment Station. Portland

Report. Pacific Southwest Forest and Range Experiment Station. Berkeley

Report of Pasture Research in the Northeastern United States. State College, Pennsylvania

Report. Pea Growing Research Organization. Yaxley

Report. Pennsylvania Agricultural Experiment Station. State College

Report. Pest Control (Officer) in Hong Kong

Report. Phytopathological Station. Ministry of Agriculture, Patras. Greece

Report. Plant Breeding Institute. Cambridge

Report. Plant Pathology Laboratory. Bvumbwe, Malawi

Report. Plant Pathology Laboratory, Ministry of Agriculture, Forests and Wildlife. Tanzania

Report. Plant Pest and Disease Situation in the Near East Region. Cairo

Report of the Potato Marketing Scheme of the Potato Marketing Board. UK

Report of the Prairie Farm Rehabilitation Service. Saskatchewan

Report of the Principal to the Governing Body. Wye College. UK

Report of Proceedings. Northern Nut Growers' Association. USA

Report of Proceedings. Western Canadian Society of Horticulture Saskatoon

Report of the Progress of Applied Chemistry. London

Report on the Progress of Chemistry London

Report. Pwllpeiran Experimental Husbandry Farm

Report. Queensland Department of Primary Industries. Brisbane

Report. Queensland Wheat Research Institute. Toowoomba

Report of the Radiation Center of Osaka Prefecture. Sakai

Report and Record of Investigations of the Department of Agriculture Uganda. Entebbe

Report and Record of Research.
East African Agriculture and
Forestry Research Organization.
Kikuyu

Report. Regional Research Centre
of the British Caribbean at the
Imperial College of Tropical
Agriculture. Trinidad

Report. Research Branch, Department of Agriculture, Sarawak.
Kuching

Report. Research Branch, Ministry
of Agriculture, Zambia. Chilanga

Report of the Research Department
of the Coffee Board. Bangalore

Report. Research Division
Department of Agriculture,
Swaziland

Report of the Research Division
Ministry of Agriculture, Sudan.
Khartoum

Report. Research Division of the
Ministry of Agriculture and
Natural Resources. Western
Region of Nigeria

Report. Research Institute
Ukiriguru. Tanzania

Report on Research and Technical
Work of the Ministry of
Agriculture for Northern
Ireland

Report. Road Research Laboratory.
Ministry of Transport.
Crowthorne

Report. Rockefeller Foundation.
New York

Report. Rocky Mountains Forest and
Range Experiment Station. Fort
Collins, Colo.

Report. Rosemaund Experimental
Husbandry Farm

Report. Rosewarne Experimental
Station and Elbridge Sub-
Station. Cambourne

Report of the Rothamsted Experimental Station. Harpenden, Herts

Report. Rubber Research Institute
of Malaysia.Kuala Lumpur

Report on Safety in Mines Research,
Ministry of Fuel and Power.
London

Report of the School of Agriculture
University of Newcastle-upon-
Tyne.

Report of the School of Agriculture
University of Nottingham.
Loughborough, Sutton Bonington

Report. Scottish Horticultural
Research Institute. Dundee

Report of the Science Service,
Department of Agriculture,
Canada. Ottawa

Report of the Scientific Research
Council, Jamaica. Kingston

Report. Scottish Plant Breeding
Station. Edinburgh

Report of the Secretary for
Agricultural Technical Services.
Republic of South Africa.
Pretoria

Report of the Secretary for
Agriculture, Rhodesia. Salsibury

Report Series. Arkansas Agricultural Experiment Station.
Fayetteville

Report. Silvicultural Research
Station, Dedza. Malawi

Report. Sleeping Sickness Service,
Northern Nigeria. Kaduma

Report. Soil Conservation Authority,
Victoria. Melbourne

Report. Soil Survey Research Board,
Agricultural Research Council.
London

Report of the Soil Survey Unit.
New South Wales Department of
Agriculture. Sydney

Report. South African Institute
for Medical Research.
Johannesburg

Report. South Carolina Agricultural Experiment Station.
Clemson

Report. South Dakota Agricultural
Experiment Station. Brookings.

Report. Southern Forest Experiment
Station. New Orleans

Report. Stockbridge Experimental Horticulture Station

Report. Sugar Cane Breeding Institute, Coimbatore. NewDelhi

Report of the Tainan Fiber Crops Experiment Station.

Report of the Taiwan Sugar Experiment Station. Taiwan

Report. Tanzania Sisal Growers' Association. Tanga

Report. Tate and Lyle Central Agricultural Research Station. Keston, Kent

Report of the Tea Institute, Taiwan

Report. Tea Research Institute, Kericho

Report of the Tea Research Institute of Ceylon, Kandy, Talawakelle

Report. Tea Scientific Department, United Planters' Association of Southern India. Coimbatore

Report. Tobacco Research Board of Rhodesia

Report of the Tobacco Research Institute. Taiwan

Report of the Tobacco Research Station. Hunsur

Report. Tocklai Experiment Station of the Indian Tea Association. Calcutta

Report. Tôhoku National Agricultural Experiment Station. Morioka

Reports of the Tottori Mycological Institute. Tottori

Report. Trawscoed Experimental Husbandry Farm

Report on Trial and Experimental Work carried out at 'Throws', Little Dunmow. UK

Report. Tropical Pesticides Research Institute. EA Common Services Organisation. Kenya

Report. Tropical Pesticides Research Institute. Tanzania

Report of the Tropical Pesticides Research Unit. Porton Down

Report. Tropical Products Institute. London

Report of the Tussock Grasslands and Mountain Lands Institute. New Zealand

Report. United Kingdom Atomic Energy Authority. Research Group. London

Report. United Nations Development Program, Technical Assistance, F.A.O. Rome

Report of the University of Nottingham, School of Agriculture

Report of the University of Sydney. School of Agriculture. Sydney

Report of the Utah Agricultural Experiment Station. Logan

Report. Veterinary Department, Kenya. Nairobi

Report of the Veterinary Department, Nigeria. Northern Region

Report of the Veterinary Department, Sierra Leone. Freetown

Report of the Veterinary Division Ministry of Agriculture, Tanzania. Dar-es-Salaam

Report of the Virginia Agricultural Experiment Station. Blacksburg

Report of the Waite Agricultural Research Institute, University of Adelaide

Report. Washington Agricultural Experiment Station. Pullman

Report. Wattle Research Institute, University of Natal. Pietermaritzburg

Report. Weed Research Organisation Agricultural Research Council. Oxford

Report. Welsh Plant Breeding Station. Aberystwyth

Report. Welsh Soils Discussion Group. Aberystwyth

Report of the West African Cocoa
Research Institute (Nigeria)
Ibadan

Report. West African Rice
Research Station. Rokupr.
Freetown

Report of the West of Scotland
Agricultural College.

Report of the Wisconsin
Agricultural Experiment Station.
Madison

Report of the Wrapper and Hookah
Tobacco Research Station.
Dinhata

Report. Wye College. Ashford
Department of Hop Research.

Report. Yundum Agricultural
Station. Gambia

Reprints. Division of Forest
Products, CSIRO, Australia.
Melbourne

Reprint. New Zealand Forest
Service, Wellington

Re: Search

Research Branch Papers. Forestry
Commission. London

Research Briefs, School of Forest
Resources, Pennsylvania State
University, University Park

Research Bulletin. Agricultural
Experimental Station, College
of Agriculture, University of
Wisconsin. Madison

Research Bulletin. Agriculture
and Home Economics Experiment
Station, Iowa State University
of Science and Technology. Ames

Research Bulletin. College of
Agriculture, University of
Missouri, Columbia, Mo.

Research Bulletin. College of
Agriculture Experiment Stations.
University of Georgia. Athens

Research Bulletin. College
Experiment Forest, Hokkaido
University, Sapporo

Research Bulletin. Department of
Agricultural Economics and
Rural Sociology. Ohio

Research Bulletin. Department of
Agricultural Economics, University
of Sydney, Sydney

Research Bulletin. Economic
Research Service, US Department
of Agriculture, Washington

Research Bulletin of the Faculty
of Agriculture, Gifu University.
Gifu

Research Bulletin. Georgia
Agricultural Experiment Station.
Athens, Ga.

Research Bulletin. Hokkaido
National Agricultural Experiment
Station. Sapporo

Research Bulletin. Idaho
Agricultural Experiment Station
Moscow

Research Bulletin of International
Center for the Improvement of
Maize and Wheat, Chapingo,
State of Mexico

Research Bulletin. Iowa
Agricultural Experiment Station.
Ames

Research Bulletin. Iowa State
University Agricultural and
Home Economics Experiment
Station. Ames

Research Bulletin. Michigan
Agricultural Experiment Station.
East Lansing

Research Bulletin. Missouri
Agricultural Experiment Station.
Columbia

Research Bulletin. Nebraska
Agricultural Experiment Station.
Lincoln

Research Bulletin of Obihiro
Zootechnical University.
Obihiro Japan

Research Bulletin. Ohio
Agricultural Experiment Station.

Research Bulletin. Ohio
Agricultural Research Development Center, Wooster. Ohio

Research Bulletin. Oregon State
University Forest Research
Laboratory, Corvallis

Research Bulletin of the Panjab
University of Science.
Hoshiarpur

Botanical Series

Research Bulletin. Production
Economic Division, Economic
Research Service, US Department
of Agriculture, Washington

Research Bulletin. Purdue University Agricultural Experiment
Station. Lafayette

Research Bulletin. Saitama
Agricultural Experiment Station.

Research Bulletin. University of
Missouri Agricultural Experiment
Station. USA

Research Bulletin. West of
Scotland College of Agriculture.
Glasgow

Research Bulletin. Wisconsin
Agricultural Experiment Station.
Madison

Research Circular. Department of
Agricultural Economics and Rural
Sociology, Ohio Agricultural
Research and Development Center.
Wooster, Ohio

Research Circular. Ohio
Agricultural Experiment Station.
Wooster

Research and Development Paper.
Forestry Commission, London

Research for Farmers. Department
of Agriculture, Canada. Ottawa

Research and Farming. North
Carolina Agricultural Experiment
Station. Raleigh

Research Investigations and Field
Trials. North of Scotland
College of Agriculture

Research Journal of the Hindi
Science Academy. Allahabad

Research Leaflet. Forest Research
Institute. New Zealand Forest
Service. Whakarewarewa

Research in the Life Sciences. USA

Research Memoirs. Empire Cotton
Growing Corporation. London

Research in the New Zealand
Department of Agriculture

Research Notes. British Columbia
Forest Service. Victoria

Research Notes. Bureau of Forestry
Manila

Research Notes. Bureau of
Resource Assessment and Land Use
Planning, University College,
Dar-es-Salaam

Research Notes. Division of Forest
Management, Forestry Commission,
N.S.W.

Research Notes. Faculty of
Forestry, University of British
Columbia. Vancouver

Research Notes. Ford Forestry
Center, L'Anse, Mich.

Research Notes. Forest Management
Research Forest Research
Laboratory. Oregon

Research Notes. Queensland Forest
Service, Brisbane

Research Notes. Texas Forest
Service. College Station

Research Notes. US Forest Service
Forest Products Laboratory.
Madison

Research Notes. US Forest Service
Intermountain Forest and Range
Experiment Station. Ogden, Utah

Research Notes. US Forest Service
Institute of Tropical Forestry
Rio Piedras. Puerto Rico

Research Notes. US Forest Service
North Central Forest Experiment
Station. St. Paul

Research Notes. US Forest Service
Northeastern Forest Experiment
Station. Upper Darby, Pa

Research Notes. US Forest Service
Northern Forest Experiment
Station, Juneau, Alaska

Research Notes. US Forest Service
Pacific Northwest Forest and
Range Experiment Station.
Portland, Ore.

Research Notes. US Forest Service
Pacific Southwest Forest and
Range Experiment Station.
Berkeley

Research Notes. US Forest Service
Rocky Mountain Forest and Range
Experiment Station. Fort Collins

Research Notes. US Forest Service
Southeastern Forest Experiment
Station. Asheville, N.C.

Research Notes. US Forest Service
Southern Forest Experiment
Station. New Orleans
Research Notes. University of
British Columbia. Forest Club.

Research Pamphlet. Forest Research
Institute, Federation of Malaysia
Kepong

Research Paper. Bureau of
Resource Assessment and Land
Use Planning, University College
Der es Salaam

Research Paper. Department of
Geography, University of
Liverpool.

Research Papers. Faculty of
Forestry, University of British
Columbia. Vancouver

Research Papers. Forest Management
Research, Forest Research
Laboratory. Oregon

Research Papers. Georgia Forest
Research Council, Macon

Research Papers. School of
Forest Resources, Pennsylvania
State University, University
Park

Research Papers. US Forest
Service, Forest Products
Laboratory. Madison

Research Papers. US Forest Service
Institute of Tropical Forestry,
Rio Piedras. Puerto Rico

Research Papers. US Forest Service
Intermountain Forest and Range
Experiment Station. Ogden, Utah

Research Papers. US Forest Service
North Central Forest Experiment
Station. St. Paul

Research Papers. US Forest Service
Northeastern Forest Experiment
Station. Upper Darby. Pa

Research Papers. US Forest Service
Northern Forest Experiment
Station, Huneau, Alaska

Research Papers. US Forest Service
Pacific Northwest Forest and
Range Experiment Station.
Portland, Ore.

Research Papers. US Forest Service
Pacific Southwest Forest and
Range Experiment Station.
Berkeley
Research Papers. US Forest Service
Rocky Mountain Forest and Range
Experiment Station. Fort Collins

Research Papers. US Forest Service
Southeastern Forest Experiment
Station. Asheville, N.C.

Research Papers. US Forest Service
Southern Forest Experiment
Station. New Orleans

Research Program Report.
Indiana Agricultural Experiment
Station, Purdue University.
Lafayette.

Research Progress Report. Purdue
University Agricultural
Experiment Station. Lafayette

Research Progress Report. Tokai-
Kinki National Agricultural
Experiment Station. Tsu City

Research Progress Report of the
Western Society of Weed Science.

Research Report. Agricultural
Economic Research Unit. Lincoln
College, Canterbury, N.Z.

Research Report. Agricultural
Experiment Station, University
of Wisconsin. Madison

Research Report. The Agricultural
Institute. Dublin

Research Report. Coffee Research
Station, Lyamungu. Tanzania

Research Report. College of
Agriculture Experiment Stations
University of Georgia. Athens

Research Report. Department of
Agriculture and Agri-business.
Louisiana State University and
Agricultural Experiment Station.
Baton Rouge

Research Report. Department of
Lands and Forests. Ontario.
Toronto

Research Report. Eastern Section
National Weed Committee. Ottawa

Research Report. Experiment
Station College of Agriculture,
University of Wisconsin

Research Report. Faculty of Agriculture and Home Economics. Univ.
ersity of Manitoba, Winnipeg

Research Report. Farmer Cooperative
Service, US Department of
Agriculture, Washington

Research Report. Federal Institute
of Industrial Research. Lagos

Research Report. Georgia
Agricultural Experiment Station.
Athens, Ga.

Research Report. Horticulture and
Forestry Division. An Foras
Taluntais (Agricultural Institute)
Dublin

Research Report. Institute of
International Agriculture Food-
Nutrition-Rural Development,
Michigan State University,
East Lansing. Mich.

Research Report. Michigan Agricultural Experiment Station.
East Lansing, Mich.

Research Report. Mississippi State
University Forest Products
Utilization Laboratory, State
College

Research Report. New Mexico
Agricultural Experiment Station.
State College

Research Report. North Carolina
Agriculture Experiment Station

Research Report. North Central
Weed Control Conference.

Research Reports of the Office of
Rural Development. Suwon, Korea

Research Report. Oregon State
University. Forest Research
Laboratory. Corvallis

Research Report. Plant Sciences and
Crop Husbandry Division. An Foras
Taluntais (The Agricultural
Institute). Dublin

Research Report. Project on the
Diffusion of Innovations in
Rural Societies, National Inst-
itute of Community Development.
Hyderabad

Research Reports of the Research
Branch. Canada Department of
Agriculture.

Research Report. Southern Weed
Conference.

Research. Timber Research and
Development Association.
High Wycombe

Research Report. United States
Army Material Command. Cold
Regions Research Engineering
Laboratory

Research Report. Western Section
National Weed Committee. Ottawa

Research Review. C.S.I.R.O.
Australia. Melbourne

Research Review. Commonwealth
Bureau of Horticulture and
Plantation Crops. East Malling

Research Series. Indian Council
of Agricultural Research

Research Studies. Washington
State University. Pullman

Research in Veterinary Science.
London, Oxford

Research and Development Report.
US Government. Springfield, Va.

Researches on Population Ecology.
Kyoto

Resenha clinico-cientifica.
São Paulo

Residue Reviews. Berlin

Residue Reviews. New York

Resource Bulletin. US Forest
Service, Intermountain Forest
and Range Experiment Station.
Ogden, Utah

Resource Bulletin. US Forest
Service, North Central Forest
Experiment Station. St. Paul

Resource Bulletin. US Forest
Service, Northern Forest
Experiment Station, Juneau,
Alaska

Resource Bulletin. US Forest
Service, Pacific Northwest
Forest and Range Experiment
Station. Portland, Ore

Resource Bulletin. US Forest
Service, Pacific Southwest
Forest and Range Experiment
Station. Berkeley

Resource Bulletin. US Forest
Service, Southeastern Forest
Experiment Station. Asheville.

Resource Bulletin. US Forest
Service, Southern Forest
Experiment Station. New Orleans

Résultats scientifiques de
l'exploration hydrobiologique
du Lac Tanzania

Results of Experiments.
Auchincruive and County Centres
West of Scotland Agricultural
College

Review of Applied Entomology.
London

Series A. Agricultural
Series B. Medical and Veterinary

Review of Applied Mycology
(new title: Review of Plant
Pathology). London

Review of the Economic Conditions
in Italy. Roma

Review of the Economic Situation
of Mexico. National Bank of
Mexico

Review of Economics and
Statistics. Cambridge, Mass.

Review of Income and Wealth.
New Haven, Conn.

Review of International
Co-operation. London

Review of Marketing and
Agricultural Economics. Sydney

Review of Medical and Veterinary
Mycology. Kew

Review of the National Research
Council. Canada. Ottawa

Review of the Polish Academy of
Sciences. Warsaw

Review of Research Work at the
Faculty of Agriculture,
Belgrade University

Review of Scientific Instruments
New York

Review of Surgery. Philadelphia

Review. Terrington Experimental
Husbandry Farm

Reviews in Engineering Geology,
Geological Society of America.
New York

Revista de la Academia colombiana
de ciencias exactas, fisicas y
naturales. Bogota

Revista agricola. Guatemala

Revista agricola. Mocambique

Revista de agricultura. Piracicaba

Revista de agricultura. Brazil

Revista de agricultura de Puerto
Rico. San Juan

Revista agronómica. Lisboa

Revista agronómica. Porto Alegre

Revista agronómica del noroeste
Argentino. S. Miguel de Tucuman

Revista Argentina de Agronomia.
Buenos Aires

Revista de la Asociación médica
argentina. Buenos Aires

Revista de Associação médica
brasileira. São Paulo

Revista da Associação médica de
Minas Gerais. Belo Horizonte

Revista de Biologia. Revista
brasileira e portuguesa de
biologia em geral. Lisboa

Revista de Biologia. Lourenco
Marques

Revista de biologia tropical.
San José. Costa Rica

Revista brasileira de biologia.
Rio de Janeiro

Revista brasileira de cirurgia.
Rio de Janeiro

Revista brasileira de economia.
Rio de Janeiro

Revista brasileira de entomologia.
São Paulo

Revista brasileira de
gastroenterologia. Rio de Janeiro

Revista brasileira de geografia
Rio de Janeiro

Revista brasileira de malariologia
e doenças tropicais. Rio de
Janeiro

Revista brasileira de medicina.
Rio de Janeiro
Revista brasileira de oftalmologia
Rio de Janeiro

Revista brasileira de tuberculose
e doenças torácicas. Rio de
Janeiro

Revista do café português. Lisboa

Revista cafetalera. Guatemala

Revista cafetera de Colombia.
Bogotá

Revista de la Catedra de
microbiologia y parasitologia.
Buenos Aires

Revista del Centro Nacional de
Patologia Animal. Peru

Revista Ceres. Vicosa

Revista chilena de higiene y
medicina preventiva. Santiago

Revista chilena de historia
natural. Santiago

Revista chilena de pediatria.
Santiago

Revista de ciências agronomicas,
(Sér.A), Louranco Marques

Revista de ciencias veterinarias.
Lisboa

Revista de cirurgia de São Paulo

Revista clinica española. Madrid

Revista clinica de São Paulo

Revista Colegio médico de
Guatemala

Revista colombiana de pediatria
y puericultura. Bogotá

Revista de la confederacion medica
panamericana. Havana

Revista dos criadores. São Paulo

Revista critica de derecho
immobiliario. Madrid

Revista cubana de ciencia
agricola. Cuba

Revista cubana de laboratorio
clinico. Habana

Revista de derecho espanol. Madrid

Revista de economia. Lisboa

Revista de economia. Mexico
Revista de economia de Galicia.
Vigo

Revista de economia politica.
Madrid

Revista ecuatoriana de entomologia
y parasitologia. Quito

Revista ecuatoriana de higiene y
medicina tropical. Guayaquil

Revista ecuatoriana de medicina y
ciencias biologicas. Quito

Revista de entomologia de
Moçambique. Lourenço Marques

Revista española de las enfermedades del aparato digestivo y
de la nutrición. Madrid

Revista española de fisiologia.
Barcelona

Revista española do lecheria.
Madrid

Revista española de oto-neuro-
oftalmologia y neurocirugia.
Valencia

Revista espanola de pediatria.
Saragoza

Revista de estudios agro-sociales.
Madrid

Revista dos estudos Gerais
Universitarios de Moçambique,
Série IV Ciências Veterinárias

Revista de faculdade de agronomia
e veterinaria da Universidade do
Rio Grande do Sul. Brazil

Revista de Faculdade de medicina
veterinaria, Universidade de
São Paulo

Revista de la Facultad de
agronomia, Universidad central
de Venezuela. Maracay

Revista de la Facultad de
agronomia, Universidad nacional
de La Plata

Revista de la Facultad de
agronomia de la Universidad de
la Republica, Uruguay. Montevide

145

Revista de la Facultad de
agronomia y veterinaria,
Universidad de Buenos Aires

Revista de la Facultad de ciencias
agrarias, Universidad nacional
de Cuyo. Mendoza

Revista de la Facultad de ciencias
medicas de Buenos Aires.

Revista de la Facultad de ciencias
médicas de la Universidad
nacional de Córdoba

Revista de la Facultad de ciencias
quimicas quimica y farmacia,
Universidad nacional de
La Plata

Revista de la facultad de ciencias
veterinarias de La Plata

Revista de la Facultad de farmacia
y bioquimica, Universidad
nacional mayor de San Marcos.
Lima

Revista de la Facultad de medicina
Mexico

Revista de la Facultad de medicina
de Tucumán. Tucumán

Revista de la Facultad de medicina
Universidad nacional de Colombia
Bogotá

Revista de la Facultad de medicina
veterinaria, Universidad
nacional mayor de San Marcos.
Lima

Revista de la Facultad de medicina
veterinaria y de zootécnia,
Universidad nacional de Colombia
Bogotá

Revista de la Facultad de medicina
veterinaria y Zootecnia.
Guatemala

Revista de la Facultad nacional de
agricultura. Bogotá

Revista de la Facultad nacional de
agronomia. Medellin

Revista forestal argentina.
Buenos Aires

Revista forestal del Peru.
La Molina

Revista forestal venezolana.
Mérida

Revista geografica del instituto
panamericano de geografia y
historia. Mexico

Revista de ginecologia e
d'obstetricia. Rio de Janeiro

Revista goiana de medicina.
Goiania

Revista de Horticultura si
Viticultura. Bucuresti

Revista do Hospital das clinicas
de Faculdade de medicina da
Universidade de São Paulo.
São Paulo

Revista del Hospital del niño.
Lima

Revista ibérica de parasitologia.
Granada

Revista industrial y agricola de
Tucumán. San Miguel

Revista do Instituto Adolfo Lutz.
São Paulo

Revista del Instituto agricola
catalán de San Isidro. Barcelona

Revista do Instituto de anti-
bioticos, Universidade do Recife
Recife

Revista Instituto Colombiano
Agropecuario

Revista do Instituto de Laticinos
Candido Tostes. Minas Gerais

Revista del Instituto Malbrán.
Buenos Aires

Revista do Instituto de Medicina
Tropical de São Paulo. São Paulo

Revista del instituto municipal
de botanica. Argentina

Revista del Instituto nacional de
biologia animal. Lima

Revista del Instituto nacional de
investigación de las ciencias
naturales y Museo argentino de
ciencias naturales 'Bernardino
Rivadavia'. Buenos Aires.

Ciencias zoológicas

Revista del Instituto de
salubridad y enfermedades
tropicales. Mexico

Revista interamericana de ciencias
agricolas.

Revista de investigaciones
agronomicas. Buenos Aires

Revista de investigaciones
agropecuarias. Buenos Aires

Ser. 1.
Ser. 2.
Ser. 3.
Ser. 4.
Ser. 5.

Revista de investigación clinica,
Hospital de enfermedades de la
nutrición. México

Revista de Investigación en Salud
Pública, Mexico

Revista de investigaciones
ganaderas. Buenos Aires

Revista Kuba de medicina tropical
y parasitologia. Habana

Revista latino-americana de
anatomia patológica. Caracas

Revista latinoamericana de
microbiologia. Mexico

Revista médica de Chile. Santiago

Revista médica de Córdoba

Revista médica dominicana.
Ciudad Trujillo

Revista médica hondureña.
Tegucigalpa

Revista médica del Hospital colonia
Mexico

Revista médica del Hospital general
Mexico

Revista médica do Sul de Minas.
Minas Gerais

Revista médica veracruzana.
Veracruz

Revista de medicina e cirurgia de
São Paulo

Revista de medicina del Estudio
general de Navarra. Pamplona

Revista de medicina experimental.
Lima

Revista de medicina pinareña.
Pinar del Rio

Revista de medicina do Rio Grande
do Sul. Pôrto Alegre

Revista de medicina veterinaria.
Buenos Aires

Revista de medicina veterinaria.
Lisboa

Revista de medicina veterinaria.
Montevideo

Revista de medicina veterinaria
y parasitologia. Maracay

Revista médico-quirúrgica de
Oriente. Santiago de Cuba

Revista mexicana de urologia.
Organo oficial de la sociedad
mexicana de urologia. Mexico

Revista del Museo de La Plata.

Seccion zoologia

Revista militar de veterinaria
Argentina

Revista de nutricion animal.
Madrid

Revista nacional de agricultura.
Bogotá

Revistă pădurilor. București

Revista del Patronato de
biologia animal. Madrid

Revista paulista de medicina.
São Paulo

Revista Portuguesa de ciencias
veterinarias. Portugal

Revista portuguesa de medicina
militar. Lisboa

Revista de sanidad e higiene
publica. Madrid

Revista de la sanidad de policia
Lima

Revista do Serviço especial de
saúde pública. Rio de Janeiro

Revista del Servicio nacional de
salud. Santiago de Chile

Revista de la Sociedad argentina
de biologia. Buenos Aires

Revista de la Sociedad colombiana
de pediatria y puericultura.
Bogota

Revista de la Sociedad entomológica argentina. Buenos Aires

Revista de la Sociedad de Medicina Veterinaria de Chile

Revista de la Sociedad mexicana de historia natural. Mexico

Revista de la Sociedad uruguaya de entomologia. Montevideo

Revista statistica. Bucureşti

Revista de Trabajo. Madrid

Revista de tuberculosis del Uruguay. Montevideo

Revista venezolana de sanidad y asistencia social. Caracas

Revista de veterinaria militar. Buenos Aires

Revista veterinaria venezolana. Caracas

Revista de veterinaria y zootecnia. Manizales, Colombia

Revista de zootehnie si medicina veterinaria. Romania

Revista di viticoltura, enologia ed agraria. Conegliano

Revue d'action populaire. Paris

Revue des agriculteurs de France. Paris

Revue de l'agriculture. Bruxelles

Revue agricole de l'Île de Réunion. Saint-Denis, Port Louis

Revue agricole et sucrière de l'Île Maurice. Port Louis

Revue algologique. Paris

Revue de biologie. Bucareşt

Revue du bois et de ses applications. Paris

Revue canadienne de biologie. Montreal

Revue de chimie. Bucureşt

Revue de chirurgie. Paris

Revue coloniale de médicine et de chirurgie. Paris

Revue des corps de santé des armées, terre, mer, air, et du Corps vétérinaire. Paris

Revue du Corps des vétérinaires biologistes des armées. France

Revue de cytologie et de cyto-physiologie végétales. Paris

Revue Défense nationale. Paris

Revue des Deux Mondes. Paris

Revue d'ecologie et de biologie du sol. Paris

Revue de l'economie politique. Paris

Revue de l'Elevage. Paris

Revue de l'Elevage. Bétail et Basse-Cour. Paris

Revue d'elevage et de medecine veterinaire des pays tropicaux. Paris

Revue de l'embouteillage et des industries connexes. Paris

Revue de la Faculté de médecine de l'Université de Téhéran

Revue des fermentations et des industries alimentaires. Gand, Bruxelles

Revue forestière française. Nancy

Revue française de l'Agriculture. Paris

Revue française d'apiculture. Paris

Revue française des corps gras. Paris

Revue française d'études cliniques et biologiques. Paris

Revue française de sociologie. Paris

Revue générale de botanique. Paris

Revue générale de botanique fondee par Gaston Bonnier. Paris

Revue générale du caoutchouc. Paris

Revue générale du caoutchouc et des plastiques

Revue générale des sciences pures et appliquées. Paris

Revue de géographie alpine. Grenoble

Revue de Géographie de Lyon. Lyon

Revue de Géographie marocaine.
Casablanca

Revue de Géographie Physique et
de Géologie Dynamique. Paris

Revue Géographique des Pyrénées
et du Sud-Ouest. Toulouse

Revue de géomorphologie dynamique.
Paris, etc.

Revue d'hématologie. Paris

Revue horticole. Paris

Revue horticole de l'Algérie,
Tunisie, Maroc. Alger

Revue horticole suisse. Genève

Revue d'hygiène et de médecine
sociale. Paris

Revue d'immunologie et de thérapie
antimicrobienne. Paris

Revue internationale d'hépatologie
Paris

Revue internationale des produits
tropicaux. Paris

Revue internationale des tabacs.
Paris

Revue juridique et économique du
Sud-Ouest (Série Économique).
Bordeaux

Revue Laitière Française
"l'Industrie Laitière". Paris

Revue du marche commun. Paris

Revue de medecine et d'hygiene
d'outre-mer. Paris

Revue de médecine vétérinaire.
Toulouse

Revue médicale de Liège

Revue médicale du Moyen-Orient.
Beyrouth

Revue médicale de la Suisse
romande. Geneve. Lousanne

Revue médicale et vétérinaire.
Paris

Revue médico-chirurgicale des
maladies du foie, du pancréas
et de la rate. Paris

Revue de mycologie. Paris

Revue de mycologie.
Supplément Colonial. Paris

Revue neurologique. Paris

Revue d'Oka, agronomie, medecine,
veterinaire. La Trappe. Quebec

Revue d'oto-neuro-ophtalmologie.
Paris

Revue de pathologie comparee
et de medicine experimentale.
Paris

Revue de pathologie vegetale et
d'entomologie agricole de
France. Paris

Revue du praticien. Paris

Revue pratique de legislation
agricole. Paris

Revue romande de biologie.
Romanie

Serie Botanique
Serie Zoologie

Revue Roumaine de Biochimie.
Bucharest

Revue Roumaine d'Inframicro-
biologie. Roumania

Revue Roumaine de Medecine
Interne. Bucharest

Revue des Sciences économiques.
Liège

Revue Socialiste. Paris

Revue de la societe belge d'etudes
et d'expansion. Liege

Revue de la Societe de biometrie
humaine. Paris

Revue suisse economique politique
et statistique. Bern

Revue suisse de Viticulture et
Arboriculture. Lausanne

Revue suisse de zoologie. Annales
de la Société zoologique suisse
et du Muséum d'histoire naturelle de Genève

Revue swisse d'agriculture

Revue trimestrielle. Association
pour la prévention de la
pollution atmospherique. Paris

Revue verviétoise d'histoire
naturelle. Verviers

Revue de zoologie agricole et
appliquée. Bordeaux, Talence

Revue de zoologie et de botanique
africaines. Bruxelles

Rheologica acta. Darmstadt

Rhode Island Agriculture.
Kingston. R.I.

Rhodesia Agricultural Journal.
Salisbury

Rhodesian Journal of Agricultural
Research. Salisbury

Rhodesian Farmer. Rhodesia

Rhodesian Journal of Economics.
Salisbury

Rhodesian Tobacco. Salisbury

Rhodesian Tobacco Journal

Rhododendron and Camellia Yearbook.
London

Rhododendron und immergrüne
Laubgehölze Jahrbuch. Deutsche
Rhododendron-Gesellschaft.
Bremen

Rhodora. Journal of the New
England Botanical Club. Boston

Ribovodstvo i Ribolovstvo

Rice Bulletin. London

Rice Journal. New Orleans

Ricerca scientifica. Roma

Ricerche economiche. Venezia

Riforma medica. Napoli

Riistatieteellisiä julkaisuja.
Helsinki

Rimba Indonesia. Bogor

Rimba muda. Bogor

Riso. Milano

Risö Report. Danish Atomic Energy
Commission Research Establishment
Risö.

Il Risparmio. Milano

Rit landbúnaðardeildar, atvinmudeih
Háskólans. Rejkjavik.

Seri. B. Flokkur

Rivista di agricoltura subtropicale
e tropicale. Firenze

Rivista di Agronomia. Italy

Rivista di anatomia patologica e
di oncologia. Padova

Rivista di biologia. Roma, Perugia

Rivista di biologia coloniale.
Roma

Rivista di clinica pediatrica.
Firenze

Rivista di diritto agrario.
Firenze

Rivista di economia agraria.
Roma

Rivista dell'Istituto sieroterapico
italiano. Napoli

Rivista italiana di economia
demografia e statistica. Roma

Rivista italiana d'igiene. Pisa

Rivista Italiana delle Sostanze
Grasse. Milan

Rivista del latte. Lodi

Rivista di malariologia. Roma

Rivista di neurologia. Napoli

Rivista dell' ortoflorofruitti-
coltura italiana. Firenze

Rivista di ostetricia e
ginecologia pratica. Palermo

Rivista di parassitologia. Roma

Rivista di patologia vegetale.
Padova.

Serie IV

Rivista di politica agraria.
Bologna

Rivista di politica economica.
Roma

Rivista di storia dell'agricoltura
Roma

Rivista di viticoltura e di
enologia. Conegliano

Rivista di zootecnia. Firenze,
Milano

Rivista di zootecnica agricoltura
veterinaria. Bologna

Riz et riziculture et cultures
vivrières tropicales.
Nogent-s.-Marne

Robigo. Castelar

Rocznik Akademii medycznej im.
Juliana Marchlewskiego w
Białystoku. Białystok

Rocznik nauk leśnych. Instytut
badawczy lesnictwa. Warszawa

Rocznik nauk rolniczych. Krakow,
Poznan, Warszawa

Seria A. Roslinne
Seria B. Zootechnika
Seria D. Monografie
Seria F. Melioracji i uzytkow
zielonych
Seria G. Ekonomia rolna
Seria H. Rybacka

Rocznik Państwowego zakładu
hygieny. Warszawa

Rocznik Sekcji dendrologicznej
Polskiego towarzystwa botanicz-
nego. Warszawa

Rocznik technologii i chemii
żywności. Warszawa

Rocznik Wyższej szkoły rolniczej
w Poznaniu. Poznań

Roczniki gleboznawcze. Warszawa

Roczniki Instytutu Przemysłu
Mleczarskiego. Warsaw

Rodo Kagaku. Tokyo

Rodriguesia. Revista do Instituto
de biologia vegetal. Rio de
Janeiro

Rohm and Haas Reporter.
Philadelphia

Rose Annual. London

Ross Laboratories Rediatric
Research Conference. Ohio

Rostlinna výroba. Ustav
vedeckotechnickych informaci
MZLVH. Praha

Rothamsted Memoirs on Agricultural
Science (Collected papers)

Round Table. London

Royal Central Asian Society
Journal. London

Royal Forest Department, Ministry
of Agriculture. Bangkok

Royal Institute of Public Health
and Hygiene Journal. UK

Royal Society of Health Journal.
London

Rozpravy Československé akademie
věd. Praha.

Řada MPV (matematických a
přirodnich věd).

Ruakura Farmers' Conference Week.

Rubber Board Bulletin. Kottayam

Rubber Developments. London

Ruch prawniczy, ekonomiczny i
socjologiczny. Poznan

Rumanian Medical Review. Bucharest

Rumanian Scientific Abstracts.
Bucharest

Natural Sciences

Rundschau fur Fleischbeschauer
und Trichinenschauer. Hannover

Rural Development. Sydney

Rural economic problems. Tokyo

Rural Life. New Barnet, UK

Rural Research in C.S.I.R.O.
Melbourne

Rural Sociology. East Lansing,
Michigan

Rural Sociology Report. Department
of Sociology and Anthropology,
Iowa State University, Ames,
Iowa

S

S.Y.S. Reporter

Saatgut-Wirtschaft. Stuttgart

Sabah Forest Record. Department
of Forestry. Jesselton

Sabouraudia. UK

Saccardoa: monographiae mycologicae
Pavia

Sadovodstvo. Moskva

Saermelding, Norges Landbruksøk-
onomiske Institut. Oslo

Saharnaja promyslennostj.

Sakharnaya svekla. Moskva

Salud publica de Mexico

Samaru Agricultural Newsletter.
Nigeria

Samaru Miscellaneous Paper.
Institute for Agricultural
Research, Ahmadu Bello University
Samaru

Samaru Research Bulletin.
Kaduna, Nigeria

Samen sterk. Netherlands

Samenfachmann

Samvirke

Sang. Biologie et pathologie.
Paris

Sanitarian. Association of Public
Health Inspectors. London

Sankhya. The Indian Journal of
Statistics. Calcutta

Santo Tomás Journal of Medicine.
University of Santo Tomás.
Manila

Sapporo Medical Journal. Sapporo

Sarawak Museum Journal. Kuching

Sarsia. Bergen

Sau og geit. Oslo

Saugetierkundliche Mitteilungen.
Germany

Savremena Poljoprivreda. Novi Sad

Sbirka zakonu.CSSR

Sbornik Československé akademie
zemédélských véd. Praha

Rada C. Rostlinna vyroba
Rada D. Lesnictvi
Rada E. Zivoćišná výroba

Sbornik Dokladov II vo Mezhoblast-
noi Konferentsii Pochvovedov i
Agrokhimikov Srednogo Povolzh'ya
i Yuzhnoi Urala. Kazanskii
Universitet. Kazan

Sbornik "Geografiya i Klassifikatsiya Pochv Azii". Moskva

Sbornik "Izmenenie Pochv pri
Okul'turivanii, ikh Klassifikatsiya i Diagnostika". Moskva

Sbornik Lékařský. Prague

Sbornik. "Mikroélementy i
Produktivnost' Pastenii". Riga

Sbornik "Mikroorganizmy v Sel'skoe
Khozyaistvo", Moskovskii
Universitet. Moskva

Sbornik narodniho musea v Praze.
Praha.
Rada B. Prirodovědný

Sbornik Nauchno-Issledovatel'-
skikh Rabot. Orlovskoi Gosudarstvennoi Sel'skokhozyaistvennoi
Opytnoi Stantsii

Sbornik nauchno-issledovatel'-
skikh rabot. Vsesoyuznogo
nauchno-issledovatel'skogo
instituta agrolesomelioratsii.

Sbornik nauchno-tekhnicheskoi
informatsii Vsesoyuznogo
instituta gelmintologii im
K.I. Skryabina

Sbornik nauchnykh rabot Altaiskoi
kraevoi nauchno-issledovatelskoi
veterinarnoi stantsii

Sbornik nauchnykh rabot. Nauchno-
issledovatel'skii Institut
Sel'skogo Khozyaistva Tsentral'-
noi-Chernozemnoi Polosy im V.V.
Dokuchaeva. Voronesh

Sbornik nauchnykh rabot Ryazan-
skogo sel'skokhozyaistvennogo
instituta

Sbornik nauchnykh rabot Sibirskogo
nauchno-issledovatel'skogo
instituta sel'skogo khozyaistva.
Omsk.

Sbornik nauchnykh rabot Sibirskogo
zonal'-nogo nauchno-issledovate-
l'skogo veterinarnogo instituta

Sbornik nauchnykh rabot Vsesoyuznyi nauchno-issledovatel'skogo
Institut Zhivotnovodstva. USSR

Sbornik nauchnykh trudov Armyanskogo sel'skokhozyaistvennogo
instituta. Erevan

Sbornik nauchnykh trudov Belorusskogo nauchno-issledovatel-
skogo veterinarnogo instituta.
Minsk

Sbornik nauchnykh trudov Donskogo
sel'skokhozyaistvennogo
instituta

Sbornik nauchnykh trudov. Estonski
nauchno-issledovatelski institut
zhivotnovodstva i veterinarii.

Sbornik nauchnykh trudov Eston-
skogo nauchnogo instituta
zemledeliya i melioratsii.
Tallin

Sbornik nauchnykh trudov Estonskoi
sel'skokhozyaistvennoi akademii.
Tallin

Sbornik nauchnykh trudov Ivanov-
skogo sel'skokhozyaistvennogo
instituta. Ivanovo

Sbornik nauchnykh trudov Push-
kinskaya nauchno-issledovatel'-
skaya laboratoriya razvedeniya
sel'skokhozyaistvennykh
zhivotnykh. USSR

Sbornik nauchnykh trudov semipalatinskogo zootekhnichesko-
veterinarnogo instituta

Sbornik nauchnykh trudov
Uzbekski nauchno-issledovatelski
veterinarni institut

Sbornik nauchnykh trudov Uzbekskoi
akademii sel'skokhozyaistvennikh
nauk

Sbornik Provozné ekonomické
ekonomicke Fakulty, Vysoká Skola
Zemědělska v Praze, Ceských,
Budejovicich

Sbornik rabot po gidrologii
leningradskogo gosudarstvennogo
gidrologicheskogo instituta

Sbornik rabot Leningradskogo
veterinarnogo instituta.
Leningrad

Sbornik rabot po lesnomu
khozyaistvu. Vsesoyuznyi nauchno-
issledovatel'skii institut
lesovodstva i mekhanizatsii
lesnogo khozyaistva. Moskva

Sbornik rabot Moskovskoi gidro-
meteorologicheskoi observatorii.
Moskva

Sbornik rabot Vologodskoi
nauchno-issledovatel'skoi
veterinarnoi opytnoi stantsii

Sbornik rabot po Zemledeliyu i
pochvovedeniyu. Saransk

Sbornik "Rohumaaviljelus". Tallin

Sbornik trudov po agronomicheskoi
fizike. Fiziko-agronomicheskii
institut Vsesoyuznoi akademii
sel'skokhozyaistvennykh nauk
imeni V.I. Lenina. Moskva

Sbornik trudov Estestvennoi
sel'skokhozyaistvennoi akademii.

Sbornik trudov Gruzinski zootekh-
nichesko-Veterinarni institut.

Sbornik trudov Khar'kovskago
veterinarnago instituta.
Khar'kov

Sbornik trudov Latviiskogo
filiala Vsesoyuznogo obshchestva
pochvovedov

Sbornik trudov nauchno-issledovatel'skii institut osnovanii i
podzemnykh sooruzhenii.
Gosstroiizdat

Sbornik trudov Povolzhskogo
lesotekhnicheskogo instituta.
Yoshkar-Ola

Sbornik trudov Yuzhnogo nauchno-
issledovatel'skogo instituta
gidrotekhniki i melioratsii.
Novocherkassk

Sbornik trudov po zashchite
pasteni.

Sbornik vědeckého lesnického
ústavu vysoké skoly zemědělské
v praze. Prague

Sbornik vědeckých praci ustředního
státniho veterináriho ustavu.
Czechoslovakia

Sbornik vedeckych praci Vyzkumneho
ustavu krmivarskeho, pohorelice.
Czechoslovakia

Sbornik "Voprosy Issledovaniya
i Ispol'zovaniya Pochv Moldavii"

Sbornik Vysoké školy chemicko-
technologiche v Praze. Praha

  Potravinářská technologie

Sbornik Vysoké školy zemedelske v
Brné. Brnó

  Rada A. Spisy Fakulty Agronomicke
  Rada B.
  Rada C. Spisy Fakulty Lesnicke
  Rada D.

Sbornik Vysoke skoly zemedelske v Praze. Praha

Sbornik Vysoke skoly zemedelske a lesnicke faculty v Brne. Brno

Rada A. Spisy fytotechnicke, zootechnicke a ekonomicke
Rada B. Spisy fakulty veterinarni
Rada D.

Sbornik Vysokej skoly pol'nohospodarskej v Nitre. Nitra

Agronomicka fakulta
Prevadzkovo-Ekonomicka fakulta

Scandinavian Journal of Clinical and Laboratory Investigation. Oslo

Scandinavian Journal of Clinical and Laboratory Investigation. Supplements. Oslo

Scandinavian Journal of Haematology Copenhagan

Scandinavian Journal of Respiratory Diseases. Denmark

Schlacht- und Viehhof-Zeitung.

Schmollers Jahrbuch für Wirtschafts und Sozialwissenschaft. Berlin

School Science Review. London

Schriften des Instituts für Städtebau und Raumordnung. Stuttgart

Schriften des Naturwissenschaftlichen Vereins für Schleswig-Holstein. Kiel

Schriften der Schweizerischen Vereinigung für Tierzucht. Bern

Schriftenreihe des AID. Land- und Hauswirtschaftlicher Auswertungs- und Informationsdienst, (AID) Bad Godesberg

Schriftenreihe des Agrarwirtschaftlichen Institut des Bundesministeriums für Land-und Forstwirtschaft. Wien

Schriftenreihe Forstliche Abterlung, Universitat Freiburg i Br.

Schriftenreihe der forstlichen Fakultät der Universität Göttingen

Schriftenreihe Institut für Landwirtschaft Betriebs- und Arbeits-Ökonomik Gundorf. Berlin

Schriftenreihe des Instituts für Soziologie an der Universität Graz. Graz

Schriftenreihe der Landesforstverwaltung, Baden-Württemberg

Schriftenreihe für ländliche Sozialfragen, Veröffentlichungen der Agrarsozialen Gesselschaft. Göttingen

Schriftenreihe des Max-Planck-Instituts für Tierzucht und Tierernährung. Mariensee

Das Schrifttum der Agrarwirtschaft. Wien

Schweinezucht und Schweinemast. Hannover

Schweizer Archiv für Neurologie und Psychiatrie. Zürich

Schweizer Archiv für Tierheilkunde. Zürich

Schweizerische Gärtnerzeitung. Zürich

Schweizerische landwirtschaftliche Zeitschrift 'Die Grüne'. Zurich

Schweizerische landwirtschaftliche Forschung. Bern-Bümpliz

Schweizerische landwirtschaftliche Monatshefte. Bern-Bümpliz

Schriftenreihe der Landwirtschaftlichen Fakultat der Universitat Kiel

Schweizerische medizinische Wochenschrift. Basel

Schweizerische Milchzeitung and le Leitier Romand. Schaffhausen

Schweizerische Zeitschrift für Forstwesen. Bern

Schweizerische Zeitschrift für Obst- u Weinbau. Frauenfeld, Wadenswil, etc.

Schweizerische Zeitschrift für Pilzkunde. Bern

Science. Baltimore, New York, etc.

Science Abstracts of China.
Peking.

Biological Sciences

Science in Agriculture. Agricultural Experiment Station.
College of Agriculture, Pennsylvania State University.
University Park

Science Bulletin of the Cotton Research Institute. Sindos

Science Bulletin. Department of Agriculture, New South Wales.
Sydney

Science Bulletin. Department of Agricultural Technical Services, Union of South Africa. Pretoria

Science Bulletin of the Faculty of Agriculture, Kyushu University. Fukuoka

Science and Culture. Calcutta

Science for the Farmer. State College, Pa.

Science Journal. London

Science Progress. London

Science Reports of the Azabu Veterinary College. Japan

Science Report of the Hyogo University of Agriculture.
Sasayama

Ser. Agricultural biology
Ser. Agricultural chemistry
Ser. Agriculture
Ser. Plant Protection

Science Reports of the Kagoshima University. Kagoshima

Science Reports of the Matsuyama Agricultural College

Science Reports of the Research Institutes, Tohoku University.
Sendai

Science Reports of the Saitama University.

Ser.B. Biology and Earth Sciences

Science Reports of the Tohoku University. Sendai

Ser. 4. Biology

Science Reports of the Yokohama National University. Kamkura

Science du sol. Annales des laboratoires G. Truffaut.
Versailles

Scientia. Bologna, Milano

Scientia Agriculturae Bohomoslovaca. Praha

Scientia silvae.

Scientia sinica. Peking

Scientific American. New York

Scientific Horticulture. Wye, etc.

Scientific Monthly. New York

Scientific Papers of the Central Research Institute, Japanese Government Monopoly Bureau.
Tokyo

Scientific Papers of the College of General Education, University of Tokyo. Tokyo

Biological Part

Scientific Papers of the Institute of Physical and Chemical Research. Tokyo

Scientific Proceedings of the Annual Meeting of the American Veterinary Medical Association.
Chicago

Scientific Proceedings of the Royal Dublin Society. Dublin

Scientific Publications. Freshwater Biological Association.
Ambleside

Scientific Reports of the Faculty of Agriculture, Ibaraki University. Ibaraki

Scientific Reports of the Faculty of Agriculture, Okayama University

Scientific Reports of the Indian Agricultural Research Institute, New Delhi. Calcutta

Scientific Reports of the Kyoto Prefectural University. Kyoto

Agriculture
Natural and Living Science

Scientific Reports of the Saikyo
University. Kyoto

Agriculture

Scientific Reports. Shiga
Agricultural College

Scientific Researches. Dacca

Scientific Survey of Porto Rico
and the Virgin Islands

Scientific and Technical Surveys.
British Food Manufacturing
Industries Research Association.
Leatherhead

Scientific Tree Topics. Stamford,
Conn.

Scientist Pakistan. Karachi

Scienza dell'Alimentazione. Milan

Scienza tecnica lattiero-casearia.
Modena

Scottish Agricultural Economics.
Edinburgh

Scottish Agriculture. Edinburgh

Scottish Farmer and Farming World
Glasgow

Scottish Forestry. Edinburgh

Scottish Geographical Magazine.
Edinburgh

Scottish Landowner. London

Scottish Medical Journal.
Edinburgh, Glasgow

Scottish Milk Marketing Board
Bulletin. Glasgow

Scottish Naturalist. Edinburgh
Scribe. Cairo

Seed Bulletin

Seed Trade Review. London

Seed World. Chicago

Seiken Ziho. Report of the
Kihara Institute for Biological
Research

Selections from China Mainland
Magazines. Hong Kong

Selektsiya i semenovodstvo.
Moskva

Sellowia. Anais botanicos do
Herbario 'Barbosa Rodrigues'.
Santa Catarina

Sel'skoe khozyaistvo kazakhstana.

Sel'skoe khozyaistvo tadshikistana.
Stalinabad

Sel'skoe khozyaistvo uzbekistana

Sel'skokhozyaistvennaya biologiya

Sel'skokhozyaistvennaya literature
SSSR

Selskostopanska misül. Sofiya

Selskostopanska nauka. Sofiya

Selskostopanska tehnika. Sofija

Semaine des hôpitaux de Paris

Semana médica. Buenos Aires

Sementi elette. Bologna, Milano

Seminar on salinity and alkali
soil problems. New Delhi

Senckenbergiana biologica.
Senckenbergische naturforschende
Gesellschaft. Frankfurt a.M.

Seria. Rynku Wiejskiego,
Spółdzielezy Institut Badawezy.
Warszawa

Separation Science. New York

Serie Agriculture. Service de
vente des publications des C.E.
Paris

Serie cientifica e tecnica.
Instituto de Algodao de
Mocambique

Série Etudes Centre International
de Hautes Etudes Agronomiques
Méditerranéennes, Institut Agronomique méditerranéen de Montpellier. Montpellier

Série Informations internes sur
l'agriculture, Communauté
économique européenne. Bruxelles

Série Travaux de Recherches.
Station d'économie rurale de
Rennes, Institut National de
la Recherche Agronomique.
Rennes

Series. Institute of Farm Income
Research. Tel Aviv

Series A.
Series B.

Series. Land Use and Production
Economics, United Nations
Special Fund, FAO Ghab Develop-
ment Project. Damascus

Service d'Exploitation Industrielle
des Tabacs et des Allumettes
(SEITA) Annales de la Direction
des Etudes et de l'equipment.
Paris

Service, Ivon Watkins. New Zealand

Service mensuel de conjoncture de
Louvain. Louvain

Shade Tree. New Brunswick. N.J.

Shandon Instrument Applications

Sheepfarming Annual. Massey
Agricultural College, Palmerston
North

Shikoku acta medica. Tokushima

Shinfield Progress

Short Term Leaflet. Ministry of
Agriculture, Fisheries and
Food. London

Sicilia sanitaria. Palermo

Silva fennica. Suomen metsatiet-
eellinen seura. Helsinki

Silvae genetica. Frankfurt a.M.

Silvical Leaflets. Bureau of
Forestry, Department of
Agriculture and Natural Resour-
ces, Philippine Islands. Manila

Silvicultura em São Paulo

Silvicultura. Uruguay

Silvicultural Notes. Ontario
Department of Lands and Forests.
Toronto
Silviculture Research Note.
Silviculture Section, Forest
Division, Lushoto, Tanzania

Sindacalismo. Roma

Singapore Medical Journal.
Singapore

Situation and Outlook Studies.
Agricultural Institute. Dublin

Sitzungsberichte der Deutschen
Akademie der Landwirtschaft-
wissenschaften zu Berlin.
Leipzig

Sitzungsberichte der Osterreich-
ischen Akademie der Wissenschaften. Wien

Mathematisch-naturwissenschaft-
liche Klasse. Abteilung I.

Sitzungsberichte der Physikalisch-
medizinischen Sozietät in
Erlangen

Skogen. Stockholm

Skogsägaren. Stockholm

Skrifter om Norsk polarinstitutt.
Oslo

Skrifter fra Økonomiske Institut.
Aarhus Universitets, Aarhus

Skrifter utgitt av det Norske
videnskapsakademi i Oslo

Skrifter utgivna av Södra Sveriges
fiskeriförening. Lund

Smithsonian Miscellaneous
Collections. Washington

Soap and Chemical Specialities.
New York

Social and Economic Administration.
Exeter

Social and Economic Studies.
Jamaica

Social Forces, Chapel Hill, North
Carolina

Social Science Information
The Hague

Socker. Handlingar. Malmö

Sociologia ruralis. Assen

Sociologia sela. Zagreb

Sociologie a Historie Zemědělstvi
Praha
Sociologus. Berlin

Soil Biology. London

Soil Biology. International Society
of Soil Science. Commission III.
Paris

Soil Biology and Biochemistry.
Oxford

Soil Bulletin. Guozi Shudian

Soil Conservation. Washington

Soil Conservation Authority,
Victoria. Melbourne

Soil and Land-Use Surveys of the
British Caribbean. Trinidad

Soil Micromorphology. Proceedings
of the Second International
Working Meeting on Soil Micromorphology, Arnhem. Amsterdam

Soil and Plant Food. Tokyo

Soil Publication. CSIRO Australia. Melbourne

Soil Science. New Brunswick, N.J. Baltimore, etc.

Soil Science and Plant Nutrition. Tokyo

Soil Survey Bulletin (Institute for Agricultural Research, Ahmadu Bells University). Samaru, Zaria, N. Nigeria

Soil Survey Papers. Netherlands Soil Survey Institute. Wageningen

Soil Survey Reports. Hokkaido National Agricultural Experiment Station. Kotoni

Soil Survey Reports. Ontario Agricultural College and Ontario Department of Agriculture. Guelph

Soil and Water. New Zealand

Soils and Fertilizers. Harpenden

Soils and Fertilizers in Taiwan Taipei

Soils and Land Use Series. Division of Soils. CSIRO, Australia. Melbourne

Soils Report. Manitoba Soil Survey. Manitoba

Sols africains. Paris, Bangni

Sonderdruck. Torfnachrichten. Bad Zwischenahn. Hannover

Sonderhefte. Bayerisches Landwirtschaftliches Jahrbuch. Munchen

Sonderheft zur Zeitschrift "Landwirtschaftliche Forschung". Darmstadt, Frankfurt a.M.

Sonderheft. Zeitschrift fur Pflanzenkrankheiten, Pflanzen-pathologie und Pflanzenschutz. Stuttgart

Soobshcheniya Akademii nauk Gruzinskoi SSR. Tbilisi

Soobshcheniya Del'nevostochnogo filiala im. V.L. Komarova Sibirskogo Otdeleniya Akademii Nauk SSSR

Soobshcheniya Instituta Lesa. Moskva

Soobshcheniya Laboratorii Agrokhimii Akademii nauk Armyanskoi SSR. Erivan

Soobshcheniya Laboratorii Lesovedeniya. Moskva

Soobshcheniya Obshchestvoi Laboratorii Agrokhimii Akademii Nauk Armyanskoi SSR.

Sotilaslääketieteellinen aikakaus-lehti. Helsinki

Sotsialisticheskoe Sel'skoe Khozyaistvo Azerbaidzhana. Baku

South African Citrus Journal

South African Forestry Journal. Pretoria

South African G.P. Review. Cape Town. (Supplement to South African Medical Journal)

South African Journal of Agricultural Science. Pretoria

South African Journal of Economics. Johannesburg

South African Journal of Laboratory and Clinical Medicine. Cape Town

South African Journal of Medical Science. Johannesburg

South African Journal of Nutrition. Cape Town. (Supplement to South African Medical Journal)

South African Journal of Obstetrics and Gynaecology. Cape Town

South African Journal of Science Cape Town

South African Medical Journal Cape Town

South African Practitioner. Johannesburg

South African Sugar Journal. Durban

South African Sugar Year Book. Durban

South Australia Journal of Agriculture. Adelaide

South Carolina Agricultural
Research. Clemson

South Dakota Farm and Home
Research. Brookings

South Indian Horticulture.
Coimbatore.

South Pacific Bulletin

Southern Co-operative Series
Bulletin. Alabama Agricultural
Experiment Station. Auburn.
Ala.

Southern Co-operative Series
Bulletin. Arkansas Agricultural
Experiment Station. Fayetteville

Southern Co-operative Series
Bulletin. Oklahoma Agricultural
Experiment Station. Stillwater

Southern Dairy Products Journal.
Atlanta.

Southern Forestry Notes.
New Orleans

Southern Lumberman. Nashville.

Southern Medical Journal.
Birmingham, USA

Southwestern Veterinarian. College
Station. Texas

Sovetskaya meditsina. Moskva

Soviet Geography. Review and
Translation. New York

Soviet Plant Physiology.
Washington

Soviet Sociology. IASP Translations
New York

Soviet studies. Oxford

Sovetskoe Gosudarstvo i Pravo.
Moskva

Sowjetwissenschaft/Gesselschafts-
wissenschaftliche Beitrage.
Berlin

Soybean Digest. Hudson, Iowa

Sozialistische Forstwirtschaft.
Berlin

Space Science Review.
Dordrecht.

Span. Shell Public Health and
Agricultural News. London

Special Bulletin of the College
of Agriculture, Utsonomiya
University. Utsonomiya

Special Bulletin. National and
University Institute of
Agriculture, Division of
Publication. Rehovot

Special Bulletin. Okayama
Prefectural Agriculture
Experiment Station. Kitagata

Special Bulletin. State of
Israel Ministry of Agriculture,
Institute of Soil Science.
Beit-Dagan

Special Publications. American
Society of Agricultural
Engineers. St. Joseph, Mich.

Special Publication. College of
Agriculture, University of
Illinois. Urbana, Ill.

Special Publications of the Taiwan
University College of
Agriculture.

Special Report to the Arctic
Institute of North America.
(Rutgers Univ.) Montreal,
New York

Special Report. Arkansas
Agricultural Experiment Station.
Fayetteville

Special Report. Cooperative Extension Service, Iowa State Univ.
of Science and Technology. Ames

Special Report. Forest Products
Research Laboratory. Ministry
of Technology. London

Special Report. Iowa State
University Agricultural and Home
Economics Experiment Station.
Ames

Special Report. Iowa State
University, Department of
Agronomy. Iowa City

Special Report. Oregon Agricultural
Experiment Station. Corvallis.
Ore

Special Report. South African
Council of Scientific and
Industrial Research. Pretoria

Special Report. Wood Research Laboratory, Virginia Polytechnic Institute, Blacksburg

Special Study. Agricultural Economics Research Council of Canada. Ottawa

Special Technical Publications. American Society for Testing Materials. Philadelphia

Specifications. South African Bureau of Standards. Pretoria

Spectrochimica acta. Berlin, London, etc.

Speculum. College of Veterinary Medicine, Ohio State University. Ohio

Speculum. Melbourne

Sperimentale. Archivio di biologia normale e patologica. Firenze

Spisy Přirodovědecké fakulty, Universita v Brne. Brno

Spolia zeylanica. Colombo Museum Colombo

Spore Newsletter. Ryde, New South Wales

Sports Turf Bulletin. St. Ives Research Station. Bingley

Spreckels Sugar Beet Bulletin Sacramento

Srpski arhiv za celokupno lekarstvo Beograd

Staat und Recht. Berlin
Staff Papers. Washington

Stain Technology. Geneva, N.Y. etc

Stal'. Moskva

Standard Specification. Standards Association of Australia. Sydney

State Forest Notes. California Division of Forestry, Sacramento

Station Bulletin. Minnesota Agricultural Experiment Station. St. Paul

Station Bulletin. Nebraska Agricultural Experiment Station. Lincoln

Station Bulletin. New Hampshire Agricultural Experiment Station. Durham

Station Bulletin. Oregon Agricultural Experiment Station. Corvallis

Station Note. Forest Wildlife and Range Experiment Station, Moscow, Idaho

Station Paper. Forest Wildlife and Range Experiment Station, Moscow, Idaho

Station Paper. United States Forest Experiment Station, Nashville, N.C.

Station Report. Horticultural Research Station, Tatura. (Victoria Department of Agriculture)

Station Technical Bulletin. Oregon Agricultural Experiment Station. Corvallis

Stations Circular. Washington Agricultural Experiment Station. Institute of Agricultural Sciences, Washington State University. Pullman

Statist. London

Statistical Bulletin. The Dairying Industry. Canberra

Statistical Bulletin. Economic Research Service, US Department of Agriculture, Washington
Statistical Bulletin of the Metropolitan Life Insurance Co. New York

Statistical Bulletin. US Department of Agriculture. Washington

Statistical News Letter

Statistical Newsletter and Abstracts. Indian Council of Agricultural Research. India

Statistical Reporter. Manila

Statistical Theory and Method. International Journal of Abstracts. Edinburgh

Statistics Section Paper. Forestry Commission. London

Statistik tidskrift. Stockholm

Statistique agricole. Paris

Statistische Praxis. Berlin

Statisztikai szemle. Budapest

Statni statky. Praha

Stato sociale. Roma

Steriods. USA

Stichting nederlands graan-centrum
Wageningen

Stiinta solului. Bucureşti

Stikstof. 's Gravenhage

Stočarstvo. Zagreb

Strahlentherapie. Berlin, Wien etc.

Studi economici. Napoli

Studi e Ricerche. Parma

Studi sassaresi. R. Università di
Sassari. Sardinia

Sezione III

Studia ekonomiczne. Warszawa

Studia entomologica. Petropolis.
Rio de Janeiro

Studia forestalia suecica.
Stockholm

Studia Helminthologica.
Czechoslovakia

Studia i Materialy Instytutu
Ekonomiki Rolnej. Warszawa

Studia Universitatis Babeş-Bolyai.
Bucureşti

Studien zur Agrarwirtschaft.
IFO-Institut für Wirtschaftsforschung. München

Studien der Europäische Wirt-
schaftsgemeinschaft, Reihe
Landwirtschaft. Brüssel

Studies in the Economics of Fruit
Farming. Department of Economics
Wye College, University of
London. Ashford

Studies from the Institute for
Medical Research, Federated
Malay States (Federation of
Malaya). Singapore

Studies. Landbouw Economische
Instituut. 's Gravenhage

Studies Official Statistics.
Ministry of Agriculture,
Fisheries and Food. London

Studies on the Soviet Union.
Munich

Studies from the Tokugawa
Institute. Tokyo

Studii şi cercetari de agronomie.
Filiala Cluj, Academia RPR.
Bucuresti

Studii şi cercetări. Baza de
cercetări ştiinţifice Timişoara,
Academia RPR. Bucureşti

Ştiinţe agricole
Ştiinţe chimice

Studii şi cercetări de biologie.
Filiala Cluj, Academia RPR.

Studii şi cercetări de biologie.
Bucuresti.

Seria "biologie animală"
Seria "biologie vegetală"

Studii şi cercetări de chimie.
Bucureşti

Studii şi cercetari economice.
Bucureşti

Studii şicercetari de inframicro-
biologie, microbiologie şi
parazitologie. Bucureşti

Studii şi cercetari Institutului
de cercetari forestiera.
Bucuresti

Studii si cercetari ştiinţifice
Filiala Iăşi, Academia RPR.
Bucuresti

Studii tehnice şi economice
Institutului Geologic al Romaniei
Ştiinţa Solului. Bucureşti

Studijni Informace Všcobecne Otázky
v Zemědělstvi. Praha

Studijni Informace Zemědělska
ekonomika. Praha

Study of Tea. Tea Division,
Tôkaikinki Agricultural
Experiment Station. Kanaya-chô

Stuttgarter Beiträge zur
Naturkunde aus dem Staatlichen
Museum für Naturkunde in
Stuttgart. Stuttgart

Subtropicheskie kul'tury. Moskva

Sudan Agricultural Journal

Sudan Journal of Administration and Development. Khartoum

Sudan Journal of Veterinary Science and Animal Husbandry. Khartoum

Sudan Medical Journal. Khartoum

Sudan Notes and Records. Khartoum

Sudan silva. Khartoum

Süddeutscher erwerbsgartner. Stuttgart

Sugar. New York

Sugar y azúcar. New York

Sugar Beet Journal. Saginaw, Mich.

Sugar Bulletin. Department of Agriculture, British Guiana. Georgetown

Sugar Bulletin. New Orleans

Sugar Journal. New Orleans

Sugar News. Manila

Sugarcane Pathologists Newsletter. Roseville

Sulphur Institute Journal. London

Šumarski List. Zagreb

Šumarstvo. Beograd

Summary Report of the Tree Nursery, Indian Head, Saskatchewan.

Sunshine State Agricultural Research Report. Gainesville

Suo. Helsinki

Suomen eläinlääkärilehti (Finsk veterinartidskrift) Finland

Suomen kemistilehti. Helsinki

Suomen laidunyhdistyksen vuosikirja. Helsinki

Suomen maataloustieteellinen seuran julkaisuja. Helsinki

Supplement. Annales de Gembloux.

Supplement. Annales de l'Institut Pasteur. Paris

Supplement to Forestry Report of the 6th Discussion Meeting. Edinburgh, Oxford

Supplement. Israel Journal of Botany. Jerusalem

Supplement Nordisk Jordbruksforskning. Oslo

Supplementum. Acta Agriculturae Scandinavica. Stockholm

Supplementum. Acta Universitatis Carolinae. Biologica. Prague

Supplementum agrokemia es talajtan

Supplementum. Annales Agriculturae Fenniae. Helsinki

Surgery. St. Louis. etc

Surgery, Gynecology and Obstetrics, Chicago

Surinaamse landbouw. Paramaribo

Survey of China Mainland Press. Peking

Süsswaren. Hamburg

Süvremenna meditsina. Scfiya

Svensk botanisk tidskrift.

Svensk frötidning. Örebro

Svensk geografisk arsbok. Lund

Svensk husdjursskötsel. Stockholm

Svensk jordbruksforskning. Årsbok. Stockholm

Svensk papperstidning. Stockholm

Svensk Trävaru- & Pappersmasse-tidning. Stockholm

Svensk valltidskrift. Sweden

Svensk veterinartidning. Sweden

Svenska läkartidningen. Stockholm

Svenska mejeritidningen. Malmö,etc

Svenska skogsvardsföreningens tidskrift. Stockholm

Sveriges skogsvardsforbunds tidskrift. Stockholm

Sveriges utsädesförenings tidskrift. Lund, etc.

Svinovodstvo. Kiev

Svinovodstvo. Moskva

Svinskotsel. Svenska svinavels-
foreningens tidskrift. Sweden

Swedish Nutrition Foundation
Symposia. Stockholm

Sydowia. Annales mycologici.
Horn.

Sylva. Edinburgh University
Forestry Society. Edinburgh

Sylva gandavensis. Gent

Sylwan. Lwów, Warszawa

Symbolae botanicae upsalienses.
Arbeten från Botaniska
institutionen i Uppsala

Symposia of the International
Society for Cell Biology.
UK/USA

Symposia of the Society for
Experimental Biology. Cambridge

Symposia of the Zoological
Society of London. London

Symposium of the British Weed
Control Council. Oxford

Symposium on Ecological Research
in Humid Tropics Vegetation
(1963) Kuching, Sarawak

Symposium Humus and Plant. Prague

Symposium over phytopharmacie.
Ghent

Symposium of the Society for
General Microbiology. London,
Cambridge, Oxford

Symposium on Water Balance of the
Soil and Forest Meteorology.
Toronto

Syrie et Monde Arabe. Damas

Systematic Zoology. Washington, etc.

Szigma. Budapest

T

TAPPI. Technical Association of
the Pulp and Paper Industry.
New York

TPI Report. Tropical Products
Institute. London

Tabacco. Roma

Tabak

Tabakpflanzer Österreichs. Linz

Tageszeitung für Brauerei. Berlin

Tagungsberichte. Deutsche Akademie
der Landwirtschaftswissenschaften
zu Berlin. Berlin

Taiwan Sugar. Taipei

Taiwania. Taipei

Tanzania Coffee News

Tappan

Tarim Bakanliği, Orman Genel
Müdürlügü Yayinlarindan.
Istanbul

Tarsadalmi szemle. Budapest

Tasmanian Journal of Agriculture.
Hobart

Tatigkeitsbericht. Arbeitsgemeinschaft zur Forderung des
Futterbaues. Switzerland

Tätigkeitsbericht. Bundesanstalt
für Pflanzenschutz. Wien

Taxicon. New York

Taxon. International Association
for Plant Taxonomy. Utrecht

Tea. Tea Boards of Kenya, Uganda
and Tanzania. Nairobi

Tea Quarterly. Tea Research
Institute Ceylon. Nuwara Eliya.
Talawakelle

Tea Research Journal. Tokyo

Teaduslike Toode Kogumik, Eesti
Maaviljeluse ja Maaparanduse
Teadusliku Uurimise Institut.
Saku

Technical Bulletin. Agricultural
Experiment Station. Oregon State
University. Corvallis

Technical Bulletin. Arizona
Agricultural Experiment Station.
Tucson

Technical Bulletin. Cocoa
Research Institute (Ghana
Academy of Sciences). Tafo

Technical Bulletin. Colorado
Agricultural Experiment Station.
Fort Collins

Technical Bulletin. Commonwealth
Institute of Biological Control.
Ottawa, Farnham Royal

Technical Bulletin. Cyprus
Department of Agriculture.
Nicosia

Technical Bulletin. Delaware
University Agricultural Experiment Station. Newark

Technical Bulletin. Department of
Agriculture, Victoria. Melbourne

Technical Bulletin. Economic
Research Service, US Department
of Agriculture, Washington

Technical Bulletin. Economic
Service Branch, Queensland
Department of Primary Industries
Brisbane

Technical Bulletin. Experimental
Forest, National Taiwan
University, Nantou

Technical Bulletin of the Faculty
of Agriculture, Kagawa University. Mikitô

Technical Bulletin. Faculty of
Agriculture, Makerere College

Technical Bulletin. Faculty of
Horticulture, Chiba University.
Matsudo

Technical Bulletin. Florida
Agricultural Experiment Station.
Gainesville

Technical Bulletin. Georgia
Agricultural Experiment Stations,
Atlanta

Technical Bulletin. Hawaii
Agricultural Experiment Station.
Honolulu

Technical Bulletin. Institute for
Land and Water Management
Research. Wageningen

Technical Bulletin. International
Rice Research Institute. Los
Banos, Laguna.

Technical Bulletin. Kansas
Agricultural Experiment Station.
Manhattan

Technical Bulletin. Maine
Agricultural Experiment Station.
Orono, Maine

Technical Bulletin. Massachusetts
Agricultural Experiment Station.
Amherst

Technical Bulletin. Michigan
Agricultural Experiment Station.
East Lansing

Technical Bulletin. Ministry of
Agriculture, Eastern Nigeria.
Enugu

Technical Bulletin. Ministry of
Agriculture, Fisheries and Food.
London

Technical Bulletin. Minnesota
Agricultural Experiment Station.
St. Paul

Technical Bulletin. Mississippi
Agricultural Experiment Station.

Technical Bulletin. Nevada
Agricultural Experiment Station.
Reno

Technical Bulletin. New Hampshire
Agricultural Experiment Station.
Durham

Technical Bulletin. New York State
Agricultural Experiment Station.
Geneva

Technical Bulletin. North Carolina
Agricultural Experiment Station.
West Raleigh

Technical Bulletin. Oklahoma
Agricultural Experiment Station.
Stillwater

Technical Bulletin. Oregon
Agricultural Experiment Station.
Corvallis

Technical Bulletin. Plant
Protection Department. Cairo

Technical Bulletin. Rhodesia
Agricultural Journal. Salisbury

Technical Bulletin. South Dakota
Agricultural Experiment Station.
Brookings

Technical Bulletin. Sulphur
Institute. Washington, London

Technical Bulletin. Taiwan
Agricultural Research Institute.
Taipei

Technical Bulletin. Taiwan
Fertilizer Company. Taipei

Technical Bulletin. US Department
of Agriculture. Washington

Technical Bulletin. US Department
of Agriculture. Agricultural
Research Service. Washington

Technical Bulletin. US Department
of Agriculture. Forest Service.
Washington

Technical Bulletin. US Department
of Agriculture. Soil Conservation
Service. Washington

Technical Bulletin. University
of Minnesota Agricultural
Experiment Station.

Technical Bulletin. Virginia
Agricultural Experiment Station.
Blacksburg

Technical Bulletin. Washington
Agricultural Experiment Stations,
Pullman

Technical Circular. Mauritius
Sugar Industry Research Institute
Reduit

Technical Communications. Bureau
of Sugar Experiment Stations,
Queensland. Brisbane

Technical Communications.
Commonwealth Bureau of Helminthology. Farnham Royal

Technical Communications.
Commonwealth Forestry Bureau.
Oxford

Technical Communications.
Department of Agricultural
Technical Services. Pretoria

Technical Communications. Royal
School of Mines. London

Technical Communications. South
Africa (Republic of) Department
of Agricultural Technical
Services. Pretoria

Technical Co-operation: A Monthly
Bibliography. UK

Technical Document. FAO Plant
Protection Committee for the
South East Asia and Pacific
Region. Bangkok

Technical Memoranda. Pea Growing
Research Organization. Yaxley

Technical Notes. Ciba - A.R.L. Ltd
Duxford

Technical Note. Department of
Forest Research. Ibadan

Technical Note. East African
Agriculture and Forestry
Research Organization. Nairobi

Technical Notes. Forest
Department, Kenya. Nairobi

Technical Notes. Forest Department
Uganda. Entebbe

Technical Notes. Forest Products
Research Institute, Philippine
Islands. Laguna

Technical Notes. Oji Institute
for Forest Tree Improvement. Oji

Technical Notes. Utilization
Section, Forest Division,
Tanzania. Moshi

Technical Paper. Adjustment Unit,
University of Newcastle upon
Tyne, Newcastle upon Tyne

Technical Paper. Agricultural
Economic Research Unit, Lincoln
College, Canterbury, N.Z.

Technical Papers. Agricultural
Experiment Station, Puerto Rico.
Rio Piedras

Technical Papers. Division of
Entomology, CSIRO Australia.
Melbourne

Technical Papers. Division of
Forest Products, CSIRO Australia
Melbourne

Technical Papers. Division of Land
Research and Regional Survey,
CSIRO Australia. Melbourne

Technical Papers. Division of
Plant Industry, CSIRO Australia,
Melbourne

Technical Papers. Forest Research
Institute, New Zealand Forest
Service. Wellington

Technical Papers. Forestry
Commission. New South Wales

Technical Paper. National Bureau
of Economic Research, New York

Technical Papers. South Pacific
Commission. Noumea

Technical Papers. University of
Puerto Rico Agricultural
Experiment Station. Rio Piedras

Technical Paper Series. Forest
Research Institute, New Zealand
Forest Service. Whakarewarewa

Technical Progress Report.
Hawaii Agricultural Experiment
Station. Honolulu

Technical Publications.
Australian Society of Dairy
Technology. Highett

Technical Publications. University
of the State of New York,
College of Forestry. Syracuse

Technical Report. Agricultural
Land Service, Ministry of
Agriculture, Fisheries and Food.
London

Technical Report. Agricultural
Research Institute. (Ghana
Academy of Sciences). Kumasi

Technical Report. British
Electrical and Allied Industries
Research Association. London

Technical Report. Central Rice
Research Institute. Cuttack

Technical Report. Faculty of
Forestry, University of
Toronto. Toronto

Technical Report of the Grassland
Research Institute. Hurley

Technical Report. School of
Forestry, North Carolina State
University. Raleigh

Technical Reports Series.
International Atomic Energy
Agency. Vienna

Technical Report. Soil Mechanics
Section. CSIRO Australia.
Melbourne

Technical Report. Soil Research
Institute. Ghana Academy of
Sciences. Kumasi
Technical Report. Texas Forest
Service. College Station

Technical Report. Weed Research
Organization. Oxford

Technical Report. Yale University
School of Forestry. New Haven

Technical Report Series. World
Health Organisation. Geneva

Technical and Scientific Papers
presented at the Annual
Conference of Manitoba
Agronomists. Canada

Technical Series. Florida State
Board of Conservation.
Tallahasee

Technicien du lait et de ses
derives. Cachan

Technique laitière. Paris

Technisch bericht. Stichting
Nederlands graan-centrum.
Wageningen

Technology in Agriculture.
London

Technology and Culture. Chicago

Technology, India. Quarterly
Bulletin of the Planning and
Development Division, Fertilizer
Corporation of India Ltd. Sindri

Technometrics. Richmond, Va.

Tecnica. Rivista de Engenharia.
Lisboa

Tecnica agricola. Associazione
dei tecnici agricoli della
provincia di Catania

Tecnica pecuaria en Mexico.
Mexico

Tecnologia. Colombia

Tegniese Medeling. Department
van Landbou-tegniese Dienste.
Pretoria

Teho. Helsinki

Tejipar. Budapest

Tejipari kutatási közlemények
Budapest

Tekel enstitüleri raporlari.
Istanbul
Teknik Forskningstiftelsen
Skogsarbeten. Stockholm

Teknik Yayinlar, Kavakçilik
Araştirma Enstutüsü, Izmit

Tekniska Högskolan Handlingar.
Stockholm

Teknisk-vetenskaplig forskning.
Stockholm

Telepulestervezesi tajekoztato.
  Budapest

Telepulestudomany i kozlemenyek.
  Budapest

Tellus. A quarterly journal of
  geophysics. Stockholm

Tennessee Farm and Home Science.

Tennessee Valley Authority.
  Division Water Control.
  Planning Division Forest
  Development

Terminology Bulletin. Food and
  Agricultural Organisation. Rome

Terra pugliese. Lecce

Terre Malgache, University of
  Madagascar. Tananarive

Terre et la vie. Paris

Terület Statisztika. Budapest

Test Memorandum. Timber Research
  and Development Association.
  London

Test Record. Timber Research and
  Development Association. London

Tetrahedron. International Journal
  of Organic Chemistry. Oxford

Tetrahedron Letters. Oxford

Texas Journal of Science.
  San Marcos

Texas Reports on Biology and
  Medicine. Galveston

Texas State Journal of Medicine.
  Austin

Textile Institute and Industry.UK

Tezisi Dokladov Nauchno-Proizvodst-
  vennoi Konferentsii po Gelmintologii v Dzhumbule

Theoretical and Applied Genetics.
  Berlin

Therapie. Paris

Therapie der Gegenwart. Berlin
  and Wien

Thorax. London

Three Banks Review. Edinburgh

Through the Leaves. Great Western
  Sugar Co. Denver

Tidskrift for Lantman. Findland

Tidsskrift for Frøavl. Kjøbenhavn

Tidsskrift for Landøkonomi.
  Kjøbenhavn

Tidsskrift for den Norske
  laegeforening Kristiana.
  Kjøbenhavn

Tidsskrift for Planteavl.
  kjøbenhavn

Tidsskrift for skogbruk.
  Kristiania

Tiedoituksia Valtion Teknillinen
  Tutkimuslaitos. The State
  Institute for Technical
  Research. Helsinki

Tiedotus, Metsäteho. Helsinki

Tierärztliche Umschau. Konstanz

Tierra. Revista de Economia
  Agraria. Bogota

Tiers-Monde. Paris

Tierzucht. Berlin

Tierzüchter. Hannover

Tijdschrift voor economische
  en sociale geographie.
  's Gravenhage, Wageningen

Tijdschrift voor entomologie
  's Gravenhage, Amsterdam

Tijdschrift van het K. Nederlandsch
  aardrijkskundig genootschap.
  Amsterdam

Tijdschrift der Nederlandsche
  heidemaatschappij. Utrecht

Timber Bulletin for Europe
  FAO Rome

Timber Development Association
  Bulletin. Wellington

Timber Grower. London

Timber Leaflet. Forest Department,
  Kenya. Nairobi

Timber and Plywood. London

Timber and Plywood Annual. London

Timber Supply Review. Melbourne,
  Canberra

Timber Trades Journal. London

Times Agricultural Review

Times Science Review. London

Tin and its Uses. Tin Research Institute. London

Tłuszcze Jadalne, Biuletyn Instytutu Przemysłu Tłuszczowego. Warsaw

Tobacco. (incl: Tobacco Science) New York

Tobacco Abstracts. North Carolina Agricultural Experiment Station. Raleigh

Tobacco Forum of Rhodesia. Salisbury

Tobacco Trust Account Digest. Australia

Tochiseidoshigaku. Tokyo

Tohoku Journal of Agricultural Research. Sendai

Tohoku Journal of Experimental Medicine. Sendai

Tokushima Journal of Experimental Medicine. Tokushima

Tokyo Jikeikai Medical Journal

Tolvmandsbladet. København

Topola. Beograd

Tórax. Montevideo

Torfnachrichten. Bad Zwischenhahn, Hannover

Torfyannaya promyshlennost'. Moskva

Tovarovedenie. Kiev

Town and Country Planning. London

Town Milk. Official Organ of the New Zealand Milk Board. Wellington

Toxicology and Applied Pharmacology New York, London

Trabajos de la Estacion Agricola Experimental de Leon. Leon

Trabajos del Jardin botánico. Universidad de Santiago. Santiago de Compostela

Trabalhos do Centro de botânico da junta de investigações du ultramar. Lisboa

Trade Yearbook. FAO Rome

Trädgårdstidningen. Jönköping

Traduction, Faculté de Foresterie et Géodésie, Université Laval. Quebec

Traeindustrien. Kobenhavn

Traktory sel'khozmashiny. Moscow

Transactions of the Agricultural Engineering Society. Tokyo

Transactions of the American Entomological Society. Philadelphia

Transactions of the American Fisheries Society. New York

Transactions. American Geophysical Union. Washington

Transactions of the American Microscopical Society. Lancaster, Pa. etc.

Transactions of the American Neurological Association. New York

Transactions of the American Society of Agricultural Engineers. Madison, Wis.

Transactions of the Bose Research Institute. Calcutta, etc.

Transactions of the British Bryological Society. Cambridge

Transactions of the British Mycological Society. London

Transactions of the Faraday Society. London

Transactions of the Hertfordshire Natural History Society and Field Club. Hertford

Transactions of the Illinois State Academy of Science. Springfield

Transactions of the Illinois State Horticultural Society. Springfield

Transactions of the Institute of British Geographers. London

Transactions of the Institution of Mining and Metallurgy. London

Transactions of the International
Congress for Agricultural
Engineering. Lausanne

Transactions of the International
Congress of Soil Science.
Bucarest

Transactions. International Soil
Conference, New Zealand. (Inter-
national Society of Soil Science)
Lower Hutt, N.Z.

Transactions of the Iowa State
Horticultural Society. Des
Moines

Transactions of the Joint Meeting
of Commissions. International
Society of Soil Science

Transactions of the Kansas
Academy of Science. Topeka

Transactions of the Kentucky
Academy of Science. Lexington

Transactions of the Lincolnshire
Naturalists' Union. Louth

Transactions. Meeting of
Commissions II and IV of the
International Society of Soil
Science. Aberdeen

Transactions of the Mycological
Society of Japan. Tokyo

Transactions of the New York
Academy of Sciences. New York

Transactions of the North American
Wildlife Conference. Washington

Transactions of the
Ophthalmological Society of
Australia. Sydney

Transactions of the
Ophthalmological Society of the
United Kingdom. London

Transactions and Proceedings of
the Botanical Society of
Edinburgh.

Transactions and Proceedings of
the Royal Society of New
Zealand. Dunedin

Transactions of the Royal
Canadian Institute. Toronto

Transactions of the Royal Society
of Canada. Ottawa

Transactions of the Royal Society
of Edinburgh. Edinburgh

Transactions of the Royal
Entomological Society of London.

Transactions of the Royal Society
of New Zealand. Dunedin

Botany
Zoology

Transactions of the Royal Society
of South Australia. Adelaide

Transactions of the Royal Society
of Tropical Medicine and Hygiene
London

Transactionsof the Shikoku
Entomological Society. Shikoku,
Matsuyama

Transactions of the Society for
British Entomology. Southampton,
etc.

Transactions of the Suffolk
Naturalists' Society. Norwich

Transactions of the Tottori
Society of Agricultural Science.
Tottori

Transactions of the Wisconsin
Academy of Sciences, Arts and
Letters. Madison

Translations. CSIRO Australia.
Melbourne

Translation. Department of
Forestry and Rural Development,
Ottawa, Canada

Translations. Faculty of Forestry
University of British Columbia.
Vancouver

Translation. Forestry Commission
London

Translation. US Forest Products
Laboratory, Madison, Wis.

Transplantation. Great Falls
Salt Lake City, Baltimore

Travaux de l'Institut de
Recherches Sahariennes. Alger

Travaux du Laboratoire forestier
de Toulouse

Travaux et Recherches de la
Faculté de Droit et de Sciences
Economiques de Paris (Séries
"Afrique"). Paris

Travaux de la Section de pédologie
de la Société des sciences
naturelles du Maroc. Rabat

Travaux de la Section scientifique
et technique, Institut français
de Pondichéry.

Travaux de la Societe des Sciences
et des Lettres de Wrocław

Travaux de la Station de
recherches (des eaux et forêts)
de Groenendael-Hoeilaart.
Groenendaal

Treaty Series. H.M. Stationery
Office. London

Tree Planters' Notes. Forest
Service. Washington

Trees Magazine. Santa Monica, Cal.

Trees in South Africa. Johannesburg

Treubia. Recueil de travaux
zoologiques, hydrobiologiques,
et océanographiques. Buitenzorg,
Bogor

Triangle. Sandoz journal of
medical sciences. Basle, etc.

Trimestre economico. Mexico

Trinidad and Tobago Forester,
Port of Spain

Tropenlandwirt. Witzenhausen a.d.
Werra

Tropical Abstracts. Amsterdam

Tropical Agriculture. The Journal
of the Imperial College of
Agriculture, Trinidad. London

Tropical Agriculture. St.Augustine

Tropical Agriculturist and
Magazine of the Ceylon Agricultural Society. Peradeniya

Tropical Agriculturist. Colombo

Tropical Diseases Bulletin. London

Tropical Ecology. Varanasi. India

Tropical and Geographical Medicine.
Amsterdam

Tropical Grasslands. Australia

Tropical Man, Department of
Anthropology, Royal Tropical
Institute. Amsterdam

Tropical Medicine. Japan

Tropical Medicine and Hygiene News.
Bethesda

Tropical Science. Tropical Products
Institute. London

Tropical Stored Products
Information. Slough

Trud i Ceni. Sofiya

Trudove naučno-izsledovatelskija
institut po Truda. Sofija

Trudy Akademii nauk Litovskoï SSR.
Vilna

Trudy Alma-Atinskogo zooveterin-
arnogo instituta. Alma Ata

Trudy Altaiskogo Sel'kokhozyaist-
vennogo Instituta

Trudy Armyanskogo Nauchno-Issledovatel'skogo Instituta Vino-
gradarstva, Vinodeliya i Plodov-
odstva. Erevan

Trudy Armyanskogo nauchno-issledovatel'skogo instituta
zhivotnovodstva i veterinarii.
Erevan

Trudy Aspirantov Gruzinskogo
Sel'skokhozyaistvennogo
instituta.

Trudy Astrakhanskogo gosudarstvennogo zapovednika

Trudy Azerbaidzhanskogo gosudarstvennogo pedagogicheskogo
instituta

Trudy Azerbaidzhanskogo nauchno-
issledovatel'skogo instituta
gidrotekhniki i melioratsii

Trudy Azerbaidzhanskogo nauchno-
issledovatel'skogo veterinarnogo
instituta

Trudy Azerbaidshanskogo politekh-
nicheskogo instituta. Baku

Trudy Azerbaidzhanskoi stantsii
Vsesoyuznogo instituta zashchity
rastenii. Leningrad

Trudy Bashkirskogo sel'skokhozyaistvennogo instituta. Ufa

Trudy Belorusskogo nauchno-issledovatel'skogo instituta
pochvovedeniya. Gorki

Trudy Belorusskoi sel'skokhozyaistvennoi Akademii. Gorki

Trudy Biologicheskogo instituta.
Sibirskoe otdelenie, Akademiya
nauk SSSR. Novosibirsk

Trudy Blagoveshchnogo sel'skokhozyaistvennogo instituta.
Blagoveshchensk

Trudy Botanicheskogo instituta.
Akademiy nauk SSSR. Leningrad,
Moskva

Seriya 1
Seriya 3
Seriya 4

Trudy Bukharskoi Oblast' Opytnoi
sel'skokhozyaistvennoi stantsii.
Tashkent

Trudy Buryat-mongol'skoi nauchno-
issledovatel'skoi veterinarnoi
opytnoi stantsii

Trudy Buryatskogo sel'skokhozyaistvennogo instituta

Trudy Chuvashskogo sel'skokhozyaistvennogo instituta.
Cheboksary

Trudy Dagestanskogo nauchno-
issledovatel'skogo instituta
sel'skogo khozyaistva

Trudy Dal'nevostochnogo nauchno-
issledovatel'skogo veterinarnogo
instituta

Trudy Darvinskogo gosudarstvennogo
zapovednika

Trudy Donskogo Zonal'nogo instituta sel'skogo khozyaistva.

Trudy Engel'gartovskoi opytno-
meliorasoi stantsii

Trudy Erevanskogo zootekhnichesko-
veterinarnogo instituta. USSR

Trudy Gel'mintologicheskoi laboratorii. Akademiya nauk SSR.Moskva.

Trudy Geograficheskogo fakul'teta
kirgizskogo universiteta.

Trudy Glavnogo botanicheskogo sada.
Akademiya nauk SSSR. Leningrad

Trudy Glavnoi geofizicheskoi
observatorii imeni A.I. Voeikova
Leningrad, Moskva

Trudy Gor'kovskogo sel'skokhozyaistvennogo instituta. Gor'kii

Trudy Gosudarstvennogo gidrologicheskogo instituta. Leningrad

Trudy Gosudarstvennogo nauchno-
kontrolnogo instituta
veterinarnykh preparatov. Moskva

Trudy Gruzinskogo nauchno-
issledovatel'skogo instituta
gidrotekhniki i melioratsii.
Tbilisi

Trudy Gruzinskogo nauchno-
issledovatel'skogo veterinarnogo
instituta

Trudy Gruzinskogo ordena trudovogo
krasnogo znameni sel'skokhozyaistvennogo instituta

Trudy Gruzinskogo sel'skokhozyaistvennogo instituta imeni L.P.
Beriya. Tbilisi

Trudy Instituta biologii. Akademiya
nauk Latviiskoi SSR. Riga

Trudy Instituta biologii. Ural'skii
filial, Akademiya nauk SSSR.
Sverdlovsk

Trudy Instituta biologii vodokhranilishch. Akademiya nauk SSSR.
Moskva

Trudy Instituta botaniki. Akademiya
nauk Kazakhskoi SSR. Alma-Ata

Trudy Instituta eksperimental'noi
biologii akademiya nauk kazakhskoi SSR. Alma-Ata

Trudy Instituta fiziologii imeni
I.P. Pavlova. Akademiya Nauk
SSSR. Moscow

Trudy Instituta genetiki. USSR

Trudy Instituta karakulevodstva
Uzbekskoi akademii sel'skokhozyaistvennykh nauk. Samarkand

Trudy Instituta lesa i drevesiny.
Akademiya nauk SSSR. Moskva,
Leningrad

Trudy Instituta lesa i drevesiny.
Sibirskoe otdelenie, Akademiya
nauk SSSR. Moskva

Trudy Instituta mikrobiologii.
Akademiya nauk Latviiskoi SSR.
Riga

Trudy Instituta morfologii zhivot-
nykh im. A.N. Severtsova.
Akademiya nauk SSSR. Moskva,
Leningrad

Trudȳ instituta pochvovedeniya i
   agrokhimii. Akademiya nauk
   Azerbaidzhanskoi SSR. Baku

Trudȳ instituta pochvovedeniya.
   Akademiya nauk kazakhskoi SSR.
   Alma-Ata

Trudȳ instituta pochvovedeniya.
   Akademiya nauk Gruzinskoi SSR.
   Tbilisi

Trudȳ instituta pochvovedeniya.
   Belorusskoi SSR. Gorki

Trudȳ instituta pochvovedenya,
   melioratsii i irrigatsii.
   Akademiya nauk Tadzhikskoi SSR.
   Stalinabad

Trudȳ instituta veterinarii.
   Akademiya nauk Kazakhskoi SSR.
   Alma-Ata

Trudȳ instituta zoologii. Akademiya
   nauk Azerbaidzhanskoi SSR. Baku

Trudȳ instituta zoologii. Akademiya
   nauk Gruzinskoi SSR. Tbilisi

Trudȳ instituta zoologii. Akademiya
   nauk Kazakhskoi SSR. Alma-Ata

Trudȳ instituta zoologii. Akademiya
   nauk Ukrainskoi SSR. Kiev

Trudȳ instituta zoologii i
   parazitologii akademiya nauk
   Kirgizskoi SSR. Frunze

Trudȳ instituta zoologii i
   parazitologii akademiya nauk
   Turkmenskoi SSR.

Trudȳ instituta zoologii i
   parazitologii akademiya nauk
   Uzbekskoi SSR. Tashkent

Trudȳ instituta zoologii i
   parazitologii im. E.N. Pavlov-
   skogo. Akademiya nauk Tadzhikskoi SSR. Stalinabad

Trudȳ kabardino-balkarskoi
   gosudarstvennoi sel'skokhozyaistvennoi opȳtnoi stantsii

Trudȳ kafedry pochvovedeniya
   biologopochvennogo fakul'teta,
   Kazakhskii gosudarstvennyi
   universitet imeni S.M. Kirova.
   Alma-Ata

Trudȳ kaliningradskoi nauchno-
   issledovatel'skoi veterinarnoi
   stantsii.

Trudȳ karadahs'koyi nauchnoyi
   stantsiyi imeni T.I. Vyazems'koho
   Simferopol'.

Trudȳ karel'skogo filiala.
   akademiya nauk SSSR. Petrozavodsk

Trudȳ kaykazskogo gusudarstvennogo zapovednika

Trudȳ kazanskogo gusudarstvennogo
   pedagogicheskogo instituta.
   Kazan'.

Trudȳ kazakhskogo sel'skokhozyaistvennogo instituta. Alma-Ata

Trudȳ kazanskogo sel'skokhozyaistvennogo instituta. Kazan'.

Trudȳ khar'kovskogo sel'skokhozyaistvennogo instituta. Khar'kov

Trudȳ kirgizskogo nauchno-issledovatel'skogo instituta
   zemledeliya. Frunze

Trudȳ kirgizskogo sel'skokhozyaistvennogo instituta. Frunze

Trudȳ kirgizskoi lesnoi opytnoi
   stantsii

Trudȳ kishinevskogo sel'skokhozyaistvennogo instituta. Kishinev

Trudȳ komi filiala akademii nauk
   SSSR. Sȳktȳvkar

Trudȳ komissii analiticheskoi
   khimii. Akademiya nauk SSSR.
   Moskva

Trudȳ konferenzii pochvovedov
   sibiri i dal'nogo vostoka

Trudȳ kostromskogo sel'skokhozy-
   stvennogo instituta. Kostroma

Trudȳ krasnoyarskogo sel'skokhozyaistvennogo instituta.
   Krasnoyarsk

Trudȳ krymskogo meditsinskogo
   instituta

Trudȳ kubanskogo sel'skokhozyaistvennogo instituta. Krasnodar

Trudȳ kuibȳshevskogo gosudarstvennogo meditsinskogo instituta

Trudȳ laboratorii evolyutsii,
   ekologii, fiziologii, institut
   fiziologii rastenii akademii
   nauk SSSR

Trudy laboratorii gidrogeologich-
eskikh problem im F.P. Savaren-
skogo. Moskva

Trudy laboratorii lesovedeniya.
akademiya nauk SSSR. Moskva

Trudy latviiskogo nauchno-issledovatel'skogo instituta
gidrotekhniki i melioratsii.
Riga

Trudy latviiskogo sel'skokhozyaistvennogo instituta. Riga

Trudy leningradskogo gidrometeorol-
ogicheskogo instituta. Leningrad

Trudy leningradskogo obshchestva
estestvoispytatelei. Leningrad,
Moscow

  Otdelenie zoologii i phiziologii

Trudy leningradskogo sanitarno-
gigienicheskogo meditsinskogo
instituta. Leningrad

Trudy litovskogo nauchno-issledovatel'skogo instituta
zemledeliya. Dolnuva

Trudy molodykh uchenii ukrainskoi
sel'skokhozyaistvennoi akademii
Kiev

Trudy moskovskoi veterinarnoi
akademii. Moskva

Trudy murmanskogo biologicheskogo
instituta. Murmanskii morskoi
biologicheskii institut,
Akademiya nauk SSSR

Trudy murmanskoi biologicheskoi
stantsii. Kil'skii filial im
S.M. Kirova, Akademiya nauk SSSR

Trudy nauchnogo instituta po
udobreniyam i insektofungitsidam
im Ya. V. Satoilova. Moskva

Trudy nauchno-issledovatel'skogo
instituta pchelovodstva. Moskva

Trudy nauchno-issledovatel'skogo
instituta pochvovedeniya
agrokhimii Armyanskoi SSR

Trudy nauchno-issledovatel'skogo
instituta pochvovedeniya
gosudarstvennyi komitet po
khlopkovodstvu srednei azii pri
gosplane SSSR. Tashkent

Trudy nauchno-issledovatel'skogo
instituta prudovogo rybnogo
khozyaistva

Trudy nauchno-issledovatel'skogo
veterinarnogo instituta. Minsk

Trudy novocherkasskogo inzherno-
meliorativnogo instituta.
novocherkassk

Trudy novocherkasskogo zootekhnic-
hesko-veterinarnogo instituta
imeni L.I. Konnoi Armii. USSR

Trudy ob''edineniya nauchnoi sessi
gruzinskogo, Azerbaidzhanskogo
i armyanskogo sel'skokhozyaistvennykh institutov. Erevan

Trudy omskogo sel'skokhozyaistvennogo instituta. Omsk

Trudy omskogo veterinarnogo
instituta. Omsk

Trudy ostashkovakogo otdeleniya
gosniorkh

Trudy permskogo sel'skokhozyaistvennogo instituta. Molotov

Trudy pervoi sibirskoi konferentsii
pochvovedov, Akademiya nauk
SSSR Sibirskoe otdelenie.
Krasnoyarsk

Trudy pochvennogo instituta imeni
V.V. Dokuchaeva. Akademiya nauk
SSSR. Moskva, Leningrad

Trudy po prikladnoi botanike,
genetike i selektsii. Leningrad

Trudy problemnykh i tematicheskikh
soveshchanii. Zoologicheskii
institut, Akademiya nauk SSSR.
Leningrad, Moskva

Trudy pushkinskoi nauchno-issledovatel'skoi laboratorii razvedeniya sel'skokhozyaistvennykh
zhivotnykh.

Trudy radiatsii i gigieny leningradskogo nauchno-issledovatel'-
skogo instituta radiatsii
gigieny. Leningrad

Trudy saratovskogo sel'skokhozyais-
vennogo instituta. Saratov

Trudy saratovskogo zoovetinstituta.

Trudy semipalatinskogo zooveterin-
arnogo instituta

Trudȳ sevastopol'skoi biologicheskoi stantsii, Akademiya nauk
SSSR

Trudȳ severnogo nauchno-issledovatel'skogo instituta
gidrotekhniki i melioratsii.
Leningrad

Trudȳ severo-osetinskogo sel'skokhozyaistvennogo instituta.
Dzandzhikar

Trudȳ severo-zapadnogo nauchno-
issledovatel'skogo instituta.
Sel'skogo khozyaistva

Trudȳ sibirskoi konferentsii
pochvovedeniya. Krasnoyarsk

Trudȳ sikhote-alinskogo gosudarstvennogo zapovednika. Moskva

Trudȳ solikamskoi sel'skokhozyaistvennoi opytnoi stantsii

Trudȳ sredne-aziatskoj opytnoj
stancii vsesojuznogo instituta
rastenievodstva

Trudȳ stalingradskogo sel'skokhozyaistvennogo instituta.
Stalingrad

Trudȳ stavropol'skogo sel'skokhozyaistvennogo instituta.
Stavrolpl'

Trudȳ sverdlovskogo sel'skokhoz-
aistvennogo instituta.
Sverdlovsk

Trudȳ tadzhikskogo nauchno-issledovatel'skogo instituta sel'skogo
instituta sel'skogo khozyaistva.

Trudȳ tashkentskogo politekhnic-
heskogo instituta. Tashkent

Trudȳ tbilisskogo gosudarstvennogo
universiteta. Tbilisi

Trudȳ tomskogo gosudarstvennogo
universiteta. Tomsk
Trudȳ troitskogo veterinarnogo
instituta. Troitsk

Trudȳ tsentral'no-chernozemnogo
gosudarstvennogo zapovednika

Trudȳ tsentral'nogo instituta
prognozov. Leningrad

Trudȳ tsentralnogo nauchno-issledovatel'skogo instituta mekhanizatsii i elektrifikatsii sel'-
skogo khozyaistva nechernozemnoi
zony SSSR

Trudȳ Tsentral'nogo sibirskogo
botanicheskogo sada

Trudȳ turkmenskogo nauchno-issledovatel'skogo instituta zemledeliya. Ashkhabad

Trudȳ turkmenskogo nauchno-issledovatel'skogo instituta
zhivotnovodstva i veterinarii

Trudȳ turkmenskoj opytnoj stancii
VIR

Trudȳ ukrainskogo gidrometeorolo-
gicheskogo instituta. Kiev

Trudȳ ukrainskogo nauchno-issledovatel'skogo gidrometeorolog-
icheskogo instituta. Kiev

Trudȳ ukrainskogo nauchno-issledovatel'skogo instituta sel'skogo
khozyaistva. Sverdlovsk

Trudȳ ukrainskogo respublikanskogo
nauchnogo obshchestva parazito-
logov.

Trudȳ ul'yanovskogo sel'skokhozyaistvennogo instituta

Trudȳ Ural'skogo nauchno-issledovatel'skogo instituta
sel'skogo khozyaistva. Sverdlovsk

Trudȳ uzbekskogo nauchno-issledovatel'skogo instituta
veterinarii

Trudȳ volgogradskoi opytno-
meliorativnoi stantsii

Trudȳ voronezhskogo gosudarstvennogo zapovednika. Moskva

Trudȳ voronozhskogo zoovetinstituta

Trudȳ vostochno-sibirskogo
filiala. Sibirskogo otdela
akademii nauk SSSR. Moskva
Trudȳ vsesoyuznogo entomologiche-
skogo obshchestva. Akademiya
nauk SSSR. Moskva

Trudȳ vsesoyuznogo gidrobiologiche-
skogo obshchestva. Moskva

Trudȳ vsesoyuznogo instituta
eksperimental'noi veterinarii.
Moskva

Trudȳ vsesoyuznogo instituta
gel'mintologii. Moskva

Trudȳ vsesoyuznogo instituta
zashchitȳ rastenii. Leningrad,
Moskva

Trudy vsesoyuznogo nauchno-issledovatel'skogo instituta gidrotekhniki i melioratsii. Moskva

Trudy vsesoyuznogo nauchno-issledovatel'skogo instituta
molochnoi promyshlennoisti.
Moscow

Trudy vsesoyuznogo nauchno-issledovatel'skogo instituta
sakharnoi svekly i sakhara

Trudy vsesoyuznogo nauchno-issledovatel'skogo instituta spirt-
ovoi promyshlennosti.

Trudy vsesoyuznogo nauchno-issledovatel'skogo instituta
sel'skokhozyaistvennoi
mikrobiologii. Leningrad

Trudy vsesoyuznogo nauchno-issledovatel'skogo instituta
torfyanoi promyshlennosti.
Moskva

Trudy vsesoyuznogo nauchno-issledovatel'skogo instituta udobre-
nii, agrotekhniki i agropoch-
vovedeniya. Moskva

Trudy vsesoyuznogo nauchno-issledovatel'skogo instituta zashchity rastenii. Moskva

Trudy vsesoyuznogo nauchno-issledovatel'skogo instituta veterinarnoi sanitarii

Trudy vsesoyuznogo nauchno-issledovatel'skogo instituta zhivotnovodstva. Moskva

Trudy zoologicheskogo instituta.
Akademiya nauk SSSR. Leningrad
Trybuna spóldzielcza. Warszawa

Tsitologiya i Genetika. USSR

T'u Jang. Guozi Shudian.

Tuatara. Journal of the Biological
Society, Victoria University
College Wellington, N.Z.

Tubercle. London

Tudomany es mezogazdasag. Budapest

Tuinbouw berichten

Tuinbouwberichten. Groningen

Tuinbouwberichten. Leuven

Tuinbouwgids. 's Gravenhage

Tuinbouwkundig onderzoek.
Jaarverslag. 's Gravenhage

Tuinbouw mededelingen 's Gravenhage

Tulane Studies in Zoology
New Orleans

Tunisie medicale. Tunis

Turang Tongbao. Nanking

Türk Askeri veteriner hekimleri
dergisi

Türk ijiyen ve tecrübi biyoloji
dergisi. Ankara

Türk tib cemiyeti mecmuasi.
Istanbul

Türk veteriner hekimleri dernegi
dergisi

Turkeys. The Journal of the
British Turkey Federation Ltd.

Turkish Bulletin of Hygiene and
Experimental Biology. Ankara

Turkish Journal of Pediatrics.
Ankara

Turrialba. Turrialba, Costa Rica

Turtox News. Chicago

Tussock Grasslands and Mountain
Lands Institute Review.
New Zealand

Two and a Bud. Tocklai Experimental
Station, Indian Tea Association.
Tocklai.

Tydskrift vir natuurwetenskappe.
South Africa

U

U.A.C.O. Timber Review. United
Africa Company Ltd. London

U and I Cultivator. Utah-Idaho
Sugar Co. Salt Lake City

UNESCO. Bulletin for Libraries

UNESCO. Humid Tropics Research

UNESCO. Proceedings of the New
Delhi Symposium on Termites
in the Humid Tropics. Paris

Uchenye zapiski Azerbaidzhanskogo
gosudarstvennogo universiteta.
Seriya Biologicheskikh Nauk.
Baku

Uchenye zapiski Azerbaidzhanskogo
sel'skokhozyaistvennogo
institute. Baku

Uchenye zapiski Azerbaidzhanskogo
universiteta. Fiziko-matematich-
eskaya khimicheskaya seriya.
Baku

Uchenye zapiski checheno-ingush-
skogo gosudarstvennogo pedagogicheskogo universiteta

Uchenye zapiski dagestanskogo
universiteta

Uchenye zapiski gorkovskogo
gosudarstvennogo pedagogicheskogo instituta.

Uchenye zapiski gorkovskogo
gosudarstvennogo universiteta.
Molotov

Uchenye zapiski kabardino-
balkarskogo gosudarstvennogo
universiteta. Nalchik

Uchenye zapiski kalininski
gosudarstvenni pedagogicheski
institut

Uchenye zapiski kazanskogo
gosudarstvennogo universiteta

Uchenye zapiski kazanskogo
veterinarnogo instituta. Kazan'.

Uchenye zapiski kuibishevski
gosudarstvenni pedagogicheski
instituta.

Uchenye zapiski kurskogo gosudarstvennogo pedagogicheskogo
instituta

Uchenye zapiski leningradskogo
gosudarstvennogo pedagogicheskogo instituta gertsena

Uchenye zapiski Leningradskogo
ordena Lenina gosudarstvennogo
universiteta im A.S. Budnova.
Leningrad

Uchenye zapiski moskovskii
gosudarstvennyi pedagogicheskii
institut imeni V.I. Lenina.
Moskva

Uchenye zapiski moskovskogo
gosudarstvennogo universiteta.
Moskva

Uchenye zapiski novgorodskogo
golovnogo pedagogicheskogo
instituta

Uchenye zapiski penzenskogo
sel'skokhozyaistvennogo
institute. Penza

Uchenye zapiski petrozavodskogo
universiteta. Petrozavod

Uchenye zapiski sel'skogo
khozyaistva dal'nogo vostoka.
Vladivostok

Uchenye zapiski Stalingradskogo
gosudarstvennogo pedagogicheskogo instituta

Uchenye zapiski tartuskogo
gosudarstvennogo universiteta.
Tartu

Uchenye zapiski vitebskogo
veterinarnogo instituta

Uchenye zapiski vologodskii
gosudarstvennyi pedagogicheskii
institut. Vologda

Uchenye zapiski yakutskogo
gosudarstvennogo universiteta

Ucetni evidence. Praha

Uebersee rundschau. Hamburg

Uganda Journal. Kampala

Ugeskrift for agronomer. Kφbenhavn

Ugeskrift for Laeger. Kjφbenhavn

Ugeskrift for landmaend.Kjφbenhavn

Ukrayinskyi biokhemichnyi Zhurnal
Kyyiv

Ukrayinskyi Khimichnyi Zhurnal.
Kiyiv

Ukrayins'kyi botanichnyi Zhurnal
Kyyiv

Ulster Medical Journal. Belfast

Umschau. Uber die Fortschritte in
Wissenschaft und Technik.
Frankfurt

Unasylva. Washington, Rome

Undersφgelser over landbrugets
Driftsforhold periodiske
Beretninger, Landφkonomiske
Driftsbureau, Kφbenhavn

Union-Agriculture. Paris

Union of Burma Journal of Life
Sciences. Rangoon

Union médicale du Canada. Montreal

United Nations Development Programme, Food and Agricultural Organization. Rome

United States Armed Forces Medical Journal. Washington

United States Atomic Energy Commission. Washington

United States Department of Agriculture. Agricultural Research Service. Washington

United States Department of Agriculture. Soil Conservation Service. Washington

United States Department of Agriculture. Soil Survey. Washington

United States Department of Agriculture (Soil Conservation Service) in Cooperation with Alabama Department of Agriculture and Industries and Alabama Agricultural Experiment Station. Washington

Universitetet i Bergen Ārbok Naturvitenskapelig rekke

University of Ankara Yearbook of the Faculty of Agriculture

University of California Publications in Botany. Berkeley

University of California Publications in Entomology. Berkeley

University of California Publications in Zoology. Berkeley

University of Durham. Department of Geography. Burham

University of Nebraska Quarterly

University of Queensland Papers of Department of Biology. Brisbane

University of Queensland Papers of Department of Botany. Brisbane

University of Washington Publications in Biology. Seattle

University of Wyoming Publications. Laramie

University of Wyoming Publications in Science

Ūrodu. Praha

Urologiya. Moskva

Uspekhi sovremennoi biologii. Moskva

Ustav Vedeckotechnickych Informaci Ministerstva Zemedelstvi. Rostlinna Vyroba. Praha

Ustav Vedeckotechnickych Informaci Ministerstva Zemedelstvi a Lesniho Hospodarstvi. Rostlinna Vyroba. Praha

Ustav Vedeckotechnickych Informaci MZLVH Studijni Informace Pudoznalstvi a Meliorace. Praha

Ustav Vedeckotechnickych Informaci Ministerstva Zemedelstvi a Vyzivy. Rostlinna Vyroba. Praha

Ustav Vedeckotechnickych Informaci Ministerstva Zemedelstvi, Lesniho a Vodnihe Hospodarstvi. Rostlinna Vyroba. Praha

Utah Farm and Home Science. Logan

Utah Science. Utah Agricultural Experiment Station. Logan

Utredning Norsk Treteknisk Institutt. Blindern. Oslo

Uttar Pradesh Agricultural University Magazine

Uzbekskii biologicheskii zhurnal Tashkent

Uzbekskii Geologicheskii zhurnal Tashkent

Uzbekskii Khimicheskii Zhurnal Tashkent

V

Vakblad voor biologien. Helder, Amsterdam

Vakblad voor de Bloemisterij.

Valosag. Budapest

Van see tot land. Zwolle

Vanasarn. Bangkok

Vår föda. Stockholm

Våra pälsdjur. Stockholm

Världshorisont. Stockholm

Varldspolitikens Dagsfragor.
Stockholm

Växt-närings-nytt. Stockholm

Växtodling. Uppsala

Växtskyddsnotiser. Statens
vaxtskyddsanstalt. Stockholm

Vědecké práce ovocnářské výzkumny
Ústav ovocnářsky v holovousich

Vědecké práce Ustredniho
Vyzkumneho ustavu zivocisne
Vyroby v Uhrinevsi

Vědecké práce Ustredniho
Vyzkumneho ustavu rastlinnej
Vyroby Piestanoch. Bratislava

Vědecké práce. Ustredniko
Vyzkumneho ustavu rostlinne
Vyroby v Praze-Ruzyni. Prague

Vědecké práce. Vyzkumneho ustavu
Bromborarskeho CSAZV v Havlick-
ove Brode

Vědecké práce. Vyzkumneho ustavu
pro Chov Hydiny v Ivanke pri
Dunaji. Czechoslovakia

Vědecké práce. Vyzkumneho ustavu
pro Chov Prasat v Kostelci nad
Orlici. Prague

Vědecké práce. Vyzkumneho ustavu
pro Chov Skotu v Rapotine.
Czechoslovakia

Vědecké práce Vyzkumneho ustavu
Krmivarskeho v Pohorelicich.

Vědecké práce Vyzkumneho ustavu
Kukurice v Trnave. Czechoslovakia

Vědecké práce Vyzkumneho ustavu
luk a pasienkov v Banskej
Bystrici. Czechoslovakia

Vědecké práce Vyskumny ustav
lesneho Hospodarstva. Banska
Stiavnica

Vědecké práce Vyzkumneho ustavu
Okrasneho Zahradnictvi v
Pruhonicich

Vědecké práce Vyzkumneho ustavu
rastlinnej vyroby v piestanoch
Vydal ustav wedeckotechnickych
informacii MPLVH v SVPL.
Czechoslovakia

Vědecké práce Vyzkumneho ustavu
rostlinne vyroby CSAZV v Praze-
Ruzyni. Czechoslovakia

Vědecké práce Vyzkumneho ustavu
veterinarni Lekarstvi v Brne

Vědecké práce Vyzkumneho ustavu
Zavlahoveho Hospodarstva v
Bratislave

Vědecké práce Vyzkumneho ustavu
Zelinarskeho CSAZV v Olomouci

Vědecké práce Vyzkumneho ustavu
Zivocisne Vyroby v Uhrinevsi.
Prague

Vědecké práce Zyskumneho ustavu
Zivocisnej Vyroby v Nitre.
Czechoslovakia

Veeteelt- en Zuivelberichten.
Netherlands

Vegetable Growers Leaflet,
National Institute of Agricultural Botany

Vegetatio. Acta Geobotanica.
Den Haag

Vegetation of Scotland. Edinburgh

Vejasarn Medical Journal. Thailand

Venture. London

Verband der Europäischen Landwirtschaft.(Confédération Européene
d l'Agriculture CEA). Brugg

Verhandlungen der Deutschen
Gesellschaft für angewandte
Entomologie. Berlin

Verhandlungen der Deutschen
zoologischen gesellschaft.
Leipzig

Verhandlungen der Internationalen
Vereinigung für theoretische und
angewandte Limnologie. Stuttgart

Verhandelingen der K. nederlandsche akademie van wetenschappen
Amsterdam

Afdeeling natuurkunde

Verhandlungen der Zoologisch-
botanischen gesellschaft in Wien

Verksamheten, Stiftelsen för
Rasförädling av Skogsträd.
Helsinki

Vermont Farm and Home Science.
Burlington

Veröffentlichungen der CEA. Paris

Veröffentlichungen. Forschungsgesellschaft für Agrarpolitik und Agrarsoziologie. Bonn

Veröffentlichungen Geobotanisches Institut. Rubel

Veröffentlichungen des Instituts für Agrarmeteorologie und des Agrarmeteorologischen Observatoriums der Karl Marx-Universität. Leipzig

Veröfftenlichung des Instituts für Genossenschaftswesen an der Philipps-Universität Marburg, Marburg a.d. Lahn

Veröffentlichungen des Instituts für Meeresforschung in Bremerhaven. Bremen

Veröffentlichungen der Landwirtschaftlich-chemischen Bundes-Versuchsanstalt. Linz

Verpackungs-Rundschau. Frankfurt-am-Main

Verslag over de Aktiviteiten. Nationaal centrum voor grasland- en groenvoederonderzoek. Belgium

Verslag van het Landbouw Economische Instituut. 's Gravenhage

Verslagen van het Centraal instituut voor landbouwkundig onderzoek. Wageningen

Verslagen van de gewone vergadering der Afdeeling natuurkunde, K. Nederlandse Akademie van wetenschappen

Verslagen van het Landbouwkundig Onderzoek in Nederland. T.N.O. Verslagen van landbouwkundige onderzoekingen (formerly van het Rijkslandbouwproefstation). 's Gravenhage

Verslagen en mededelingen. Commissie voor hydrologisch onderzoek, T.N.O. 's Gravenhage

Verslagen en mededeelingen van den Plantenziektenkundigen dienst te Wageningen

Verslagen. Meerjarenplan voor onderzoek Akkerbouwpeulvruchten. Netherlands

Verslagen. Peulvruchten Studie Combinatie. Netherlands

Versuchsergebnisse der Bundesanstalt für alpenländische Landwirtschaft. Gumpenstein

Vertragssystem. Berlin

Vestnik Akademii nauk Kazakhskoi SSR. Alma-Ata

Vestnik Akademii nauk SSSR. Moskva

Věstnik Československé společnosti zoologické. Praha

Vestnik dermatologii i venerologii. Moskva

Vestnik Karakalpakskogo filiali Akademii nauk Uzbek SSR

Vestnik khirurgii imeni I.I. Grekova. Moskva, Leningrad

Vestnik Kievs'kogo Universiteta. Kiev.

   Seriya Biologii

Vestnik Leningradskogo gosudarstvennogo universiteta. Leningrad.

   Seriya Biologii

Vestnik Leningradskogo Instituta. Leningrad

Vestnik Leningradskogo universiteta. Leningrad.

   Seriya Biologii
   Seriya Biologiya

Vestnik Moskovskogo gosudarstvennogo universiteta. Moskva

Vestnik Moskovskogo instituta geografii. Moskva

Vestnik Moskovskogo universiteta. Moskva

   Seriya Biologii
   Seriya Pocvovedenija
   Seriya Geografii
   Seriya Geologii

Vestnik oftalmologii. Kiev, Odessa, Moskva

Vestnik rentgenologii i radiologii Leningrad, Moskva

Vestnik sel'skokhozyaistvennoi nauki. Alma-Ata

Vestnik sel'sko-khozyaistvennoi nauki. Moskva

Vestnik statistiki. Moskva

Vestnik Vyzkumnych ustavu zemedelskych

Vestsi Akademii navuk Belaruskai SSR. Minsk

Veterinär-medizinische Nachrichten Marburg

Veterinaria. Italy

Veterinaria. Madrid

Veterinaria. Revista dos alunos da Escola nacional de veterinaria. Rio de Janeiro

Veterinaria. Sarajevo

Veterinaria Colombiana. Colombia

Veterinaria italiana. Teramo

Veterinaria. Rassegna di informazione e aggiornamento

Veterinaria Spofa. Czechoslovakia

Veterinaria. Zbornik radova iz oblasti animalne proizvodnje.

Veterinaria y zootecnia. Lima

Veterinarian. Indianapolis

Veterinarian. London

Veterinariya. Moskva

Veterinariya. Ukraine

Veterinarna sbirka. Sofiya

Veterinarni medicina. Praha

Veterinarno-meditsinski nauki. Bulgaria

Veterinarski arhiv. Zagreb
Veterinarski glasnik. Beograd, Zagreb

Veterinársky časopis. Bratislava

Veterinářstvi. Brno, Praha

Veterinary Annual. Bristol

Veterinary Bulletin. Weybridge

Veterinary Inspector. New South Wales

Veterinary Medical Journal. Egypt

Veterinary Medical Review.
 /English translation of Veterinär-medizinische nachrichten/

Veterinary Medicine and Small Animal Clinician. USA

Veterinary Record. London

Veterinary Review. UK

Veteriner fakültesi dergisi, Ankara üniversitesi. Ankara

Veteriner fakültesi yayinlari, Ankara üniversitesi. Ankara

Vetserum. Zagreb

Victorian Horticulture Digest

Victorian Veterinary Proceedings. Australia

Victorian Yearbook. Melbourne

Vida agrícola. Peru

Videnskabelige Meddelelser fra Dansk naturhistorisk Forening i Kjøbenhavn

Vie médicale. Paris

Vie et milieu. Bulletin du Laboratoire Arago, Université de Paris

Vie et science economique. Paris

Vijnana Parishad anusandhan patrika. Allahabad

Vierteljahresberichte. Forschungsinstitut der Friedrich-Ebert-Stiftung. Bad Godesberg

Vierteljahrsschrift der Naturforschenden gessellschaft in Zürich.

Vierteljahrshefte zur Wirtschaftsforschung. Berlin

Viewpoints in Biology. UK
Vinea et vino portugaliae documenta.

Vinodelie i vinogradarstvo SSSR. Moskva

Viola. Sweden

Virchows Archiv für pathologische Anatomie und Physiologie und für klinische Medizin. Berlin

Virginia Agricultural Economics. Blacksburg, Va

Virginia Journal of Science. Richmond

Virginia Medical Monthly. Richmond

180

Virginia Veterinarian

Virology. Baltimore, New York

Visnyk Botanichnoho sadu.
Akademiya nauk Ukrayinskoyi RSR.
Kyyiv

Visnyk si'ls'ko-hospodars'koyi
nauky. Kyyiv, Kharkiv

Vita italiana. Roma

Vitamins. Japan Vitamin Society.
Kyoto

Vitamins and Hormones. New York

Vitis. Berichte über Rebens-
forschung. Geilweilerhop,
Landau/Pfalz

Vizgazdálkodás. Budapest

Vlaams diergeneeskundig
tijdschrift. Antwerpen, Gent

Vlugschriften van de Directie van
den landbouw. 's Gravenhage

Vlugschriften. Plantenziekten-
kundige dienst te Wageningen

Vodni hospodářstvi. Praha

Vodohospodarsky casopis. Bratislava

Voeding. Den Haag

Voeding and Techniek. Doetinchem.
The Netherlands

Vojno-sanitetski pregled. Beograd

Volksopvoeding. Amsterdam

Voprosy ekonomiki. Moskva

Voprosy fiziologii i biokhimii
kulturnykh rastenii Akademiya
nauk Moldavskogo SSR. Kishiven

Voprosy Genesisa i Krypnomashtab-
noi Kartirovanii Pochv. Kazanskii
Universitet

Voprosy geografii. Moskva

Voprosy ikhtiologii. Otdelenie
biologicheskikh nauk. Akademiya
nauk SSSR. Moskva

Voprosy istorii K.P.S.S. Moskva

Voprosy mikrobiologii, Akademiya
nauk Armyanskoi SSR

Voprosy neurokhirurgii. Moskva

Voprosy pitaniya. Moskva

Voprosy sovremennoi fiziki i
matematiki, Akademiya nauk
Uzbekskoi SSR. Tashkent

Voprosy virusologii. Moskva

Vox sanguinis. Amsterdam, Basel

Vysoká Škola. Praha

Vystavba socialisticke vesnice.
Praha

Výživa lidu. Praha

Výživa a Zdravie. Bratislava

W

WHO Chronicle. World Health
Organization. Geneva, New York

Wakayama Medical Reports.
Wakayama

Wald und Holz. Solothurn,
Switzerland

Waldhygiene. Würzburg

Wallerstein Laboratories
Communications. New York

War on Hunger. Washington

Warta Penelitian Pertanian

Washington Research Progress

Wasmann Journal of Biology.
San Francisco

Wasser und Nahrung

Wasserwirtschaft und Wassertechnik
Berlin

Water Resources Research.
Washington

Water and Sewage Works. Chicago,
New York

Water-Supply and Irrigation Paper.
Geological Survey. Washington

Watsonia. Journal of the
Botanical Society of the
British Isles. Arbroath

Weather. Royal Meteorological
Society. London

Webbia. Racoolta di scritti
botanici. Firenze

Weed Abstracts. Oxford, London

Weed Research. UK

Weed Science. Urbana, New York

Weekblad voor bloembollencultuur.
Haarlem, etc.

Weibulls allehanda. Landskrona

Weibulls Årsbok for Vaxtforadling
och Vaxtodling. Sweden

Wein-Wissenschaft. Mainz. Berlin

Wein-Wissenschaft. Beilage zur
Fachzeitschrfit "Der Deutsche
Weinbau". Berlin, Mainz

Weinberg und Keller. Frankfurt a.M

Weltwirtschaft. Kiel

Weltwirtschaftliches Archiv.
Hamburg

West African Journal of Biological
and Applied Chemistry. London

West African Journal of Biological
Chemistry. Ibadan, London

West African Medical Journal.
Lagos, Ibadan, etc.

West African Pharmacist. Kumasi,
London

West Indian Medical Journal.
Kingston, St. Andrew

West Pakistan Journal of
Agricultural Research

West Pakistan Journal of
Agricultural Science. Lahore

West Virginia Agriculture and
Forestry

Western Economic Journal.
Salt Lake City, Utah

Western Fruit Grower. San Francisco

Western Journal of Surgery,
Obstetrics and Gynecology.
Portland, Ore.

Western Political Quarterly,
Salt Lake City, Utah

Westminster Bank Review. London

Wetenschappelijke mededeelingen
K. nederlandse natuurhistorische
vereniging. Amsterdam

Weterynaria. Poland

Weyerhaeuser Forestry Paper
Centralia. Washington

Wheat Information Service.
Laboratory of Genetics, Biologic‑
al Institute, Kyoto University

Wiadomości Instytutu Melioracji
i Uzytków Zielonych. Poland

Wiadomości lekarskie. Warszawa

Wiadomości parazytologiczne.
Warszawa

Wiadomości tytoniowe. Warszawa

Wiener klinische Wochenschrift.
Wien

Wiener medizinische Wochenschrift
Wien

Wiener tierärztliche Monatsschrift
Wien, Leipzig

Wiener Zeitschrift für innere
Medizin und ihre Grenzgebiete.
Wien

Wiener Zeitschrift für Nervenheil-
kunde und deren Grenzgebiete.
Wien

Wies wspolczesna. Warszawa

Wildlife Diseases. Washington

Wildlife Monographs. Wildlife
Society. Chestertown, etc.

Wildlife Review

Wilhelm Roux Archiv für Entwick-
lungs-mechanik der Organismen.
Leipzig

Willdenowia. Mitteilungen aus dem
Botanischen Garten und Museum
Berlin-Dahlem. Berlin

William L. Hutcheson Memorial
Forest Bulletin. New Brunswick

Window. Tokyo

Wirtschaft. Berlin

Wirtschaftsdienst. Hamburg

Wirtschaftseigene Futter. Germany

Wirtschaftspolitische Mitteilungen
Zürich

Wirtschaftswissenschaft. Berlin

Wirtschaftswissenschaftliches
Mitteilungen. Köln

Wisconsin Conservation Bulletin.
Madison

Wisconsin Medical Journal.
Milwaukee. Madison

Wissenschaft und Technik der
socialistischer landwirtschaft.
Halle

Wissenschaftliche Abhandlungen
der Deutschen Akademie der
Landwirtschaftswissenschaften
zu Berlin. Leipzig

Wissenschaftliche Arbeiten
Landwirtschaftliche Hochschule
"Wassil Kalarow". Plovdiv:
Fakultat fur Ackerbau

Wissenschaftliche Beiträge
Hochschule ZK SED. Berlin

Wissenschaftliche Schriftenreihe
des Bundesministeriums für
wirtschaftliche zusammenarbeit.
Stuttgart

Wissenschaftlich-technischer
Fortschritt für die Landwirtschaf

Wissenschaftliche Zeitschrift
der Ernst Moritz Arndt-Universit-
ät Greifswald. Greifswald

Wissenschaftliche Zeitschrift der
Friedrich Schiller-Universität,
Jena. Jena

Wissenschaftliche Zeitschrift der
Hochschule für landwirtschaftliche Produktions-Genossenschaf-
ten. Meissen

Wissenschaftliche Zeitschrift der
Karl-Marx-Universität Leipzig.
Leipzig

Mathematisch-Naturwissen-
schaftliche Reihe

Wissenschaftliche Zeitschrift der
Humboldt- Universität Berlin.
Berlin

Mathematisch-naturwissenschaft-
liche Reihe

Wissenschaftliche Zeitschrift der
Martin-Luther-Universitat, Halle-
Wittenberg

Mathematisch-wissenschaftliche
Reihe

Wissenschaftliche Zeitschrift der
Technischen Universität Dresden.
Dresden

Wissenschaftliche Zeitschrift des
Universitäts Halle. Halle-
Wittenberg

Wissenschaftliche Zeitschrift der
Universität Rostock. Rostock

Reihe Mathematik und Naturwissenschaften

Worcestershire Agricultural
Chronicle. Worcester

Wood. London

Wood Industry. Tokyo

Wood News. Kampala

Wood Preserving. Washington

Wood Research. Wood Research
Institute, Kyoto University.
Kyoto

Wood Science and Technology.
New York

Wood and Wood Products. Chicago

Woodlands Papers, Pulp and Paper
Research Institute of Canada.
Pointe Claire, Quebec

Woodland Research Notes. Union
Camp Corp., Georgia. Savannah

Woodlands Section Index. Canadian
Pulp and Paper Association.
Montreal

Woodworking Digest. Wheaton, Ill.

Woodworking Industry. London

Wool Economic Research Report.
Bureau of Agricultural Economics
Canberra

Wool Technology and Sheep Breeding
Sydney

Works of the Committee on
Sociology, Krakow Branch of the
Polish Academy of Sciences.
Wrocław

Work Progress Report of Crop
Improvement Mission. Vietnam

Working Paper. Corsortium for the
Study of Nigerian Rural Develop-
ment, Michigan State University.
East Lansing, Mich.

Working Paper. United Nations
Research Institute for Social
Development. Geneva

World Agricultural Economics and
Rural Sociology Abstracts

World Agricultural Production and
Trade. Statistical Report. USA

World Agriculture. Washington

World Crops. London

World Farming

World Health. WHO. Geneva

World Health Organization.
Technical Report Series

World Medical Journal. New York

World Neurology. Minneapolis

World Review of Animal Production.

World Review of Nutrition and
Dietetics. London

World Review of Pest Control.
London

World Wood. Portland, Ore.

World's Poultry Science Journal.
Ithaca, N.Y.

Wydawnictwa własne. Instytut
zootechniki. Poland

Wydawnictwo Łodzkiego Towarzystwa
Naukowego

X

X-Ray Focus. London

Y

Yale Journal of Biology and
Medicine. New Haven

Yalova-bahce Kültürleri araştirma
ve egitim merkezi dergisi.
Yalova

Yearbook of Agriculture.
US Department of Agriculture,
Washington

Yearbook of the Association of
Pacific Coast Geographers. USA

Yearbook. British Goat Society.
London

Yearbook. California Avocado
Society. Los Angeles

Yearbook of the Carnegie
Institution of Washington

Yearbook of the Faculty of
Agriculture. University of
Ankara. Ankara

Yearbook. Institute of Inspectors
of Stock, New South Wales.
Sydney

Yearbook of the Institute of
Nutrition. Budapest

Yearbook. National Chrysanthemum
Society. London

Yearbook. Royal Veterinary and
Agricultural College. Copenhagen

Yearbook. South West Africa
Karakul Breeders' Association.
Windhoek

Yojana. New Delhi

Yokohama Medical Bulletin.
Yokohama

Yonago acta medica. Yonago

Yorkshire Bulletin of Economic
and Social Research. Hull

Yonsei Business Review. Seoul

Yrkesfruktyrking. Oslo

Yugoslav Society of Soil Science
Publication. Beograd

Yugoslav Survey. Beograd

Z

Za socialistické zemědélstvi.
Praha

Za sotsialisticheskuyu sel'skokh-
ozyaistvennuyu nauku. v Praze

Zaadbelangen. 's Gravenhage

Zagadnienia ekonomiki rolnej.
Warszawa

Zalacenia Ogólne. Institut
Ochrony Roślin. Poznan

Zapiski Leningradskogo sel'skokhozyaistvennogo instituta.
Leningrad

Zapiski Voronezhskogo sel'skokhozyaistvennogo instituta.
Voronezh

Zashchita rastenii ot vreditelei
i boleznei. Moskva

Zashchitnye Lesorazvedenie.
Volgograd

Zaštita bilja. Beograd

Zbirnȳk prats Zoologichnogo
Muzeyo. Institut zoologii,
Akademiya nauk Ukrainskoi SSR

Zbornik Biotehniske Fakultete
Univerze v Ljubljani.
Ljubljana

Zbornik Instituta za gozdno in
lesno gospodarstvo slovenije.
Ljubljana

Zbornik Matice srpske. Novi Sad.

Ser. Prirod. Nauka

Zbornik radova. Bioloski institut
NR Srbije. Beograd

Zbornik radova. Institut za
Poljoprivredna Istrazivanja.
Novi Sad

Zbornik radova Poljoprivrednog
fakulteta, Universitet u
Beogradu. Beograd

Zdravstveni vestnik. Ljubljana

Zeiss-Mitteilungen. Oberkochen

Zeitschrift für Acker- und
Pflanzenbau. Berlin

Zeitschrift für Agrarokonomik.
Berlin

Zeitschrift für allgemeine Mikrobiologie. Berlin

Zeitschrift für Altenrsforschung
Dresden, Leipzig

Zeitschrift für analytische Chemie.
Wiesbaden, etc.

Zeitschrift für angewandte
Entomologie. Berlin

Zeitschrift für angewandte
Zoologie. Berlin

Zeitschrift für arztliche fort-
bildung

Zeitschrift für auslandische
Landwirtschaft. Frankfurt a.M.

Zeitschrift für Botanik. Jena

Zeitschrift für Chemie. Leipzig

Zeitschrift für Ernährungs-
wissenschaft. Darmstadt

Zeitschrift für Fischerei und
deren Hilfswissenschaften.
Berlin, etc.

Zeitschrift für Geomorphologie.
Leipzig, Berlin

Zeitschrift für die gesamte
Anatomie.

Abt. 1 Zeitschrift für Anatomie
u. Entwicklungsgeschichte

Zeitschrift für die gesamte
experimentelle Medizin. Berlin

Zeitschrift für das gesamte
Genossenschaftswesen. Gottingen

Zeitschrift für die gesamte
Hygiene und ihre Grenzgebiete.
Berlin

Zeitschrift für die gesamte
innere Medizin und ihre
Grenzgebiete. Leipzig

Zeitschrift für die gesamte
Staatswissenschaft. Tübingen

Zeitschrift für Haut- und
Geschlechtskrankheiten und
deren Grenzgebiete. Berlin

Zeitschrift für Hygiene und
Infektionskrankheiten medizinische mikrobiologie, immunologie
und virologie. Leipzig, Berlin

Zeitschrift für Immunitätsfors-
chung, Allergie und Klinische
Immunologie. Jena

Zeitschrift für Immunitätsfors-
chung und experimentelle
therapie. Jena

Zeitschrift für Infektionskrankheiten, parasitäre Krankheiten
u. Hygiene der Haustiere. Berlin

Zeitschrift für Jagdwissenschaft.
Hamburg

Zeitschrift für Kinderheilkunde.
Berlin

Zeitschrift für klinische chemie
und klinische biochemie. Berlin

Zeitschrift für Krebsforschung.
Berlin

Zeitschrift für Kreislaufforschung
Dresden, Darmstadt

Zeitschrift für Kulturtechnik und
Flurbereinigung. Berlin

Zeitschrift für Landeskultur.
Germany

Zeitschrift für landwirtschaftliches Versuchs- und Unter-
suchungswesen. Berlin

Zeitschrift für Lebensmittel-
Untersuchung und Forschung.
Munich

Zeitschrift für Medizinische
Mikrobiologie und Immunologie

Zeitschrift für mikroskopisch-
anatomische Forschung. Leipzig

Zeitschrift für Morphologie und
Ökologie der Tiere. Berlin

Zeitschrift für Nationalökonomie.
Wien

Zeitschrift für Naturforschung.
Wiesbaden. Tübingen

B. Chemie, Biochemie, Biophysik,
Biologie und verwandte gebiete

Zeitschrift für Parasitenkunde.
Berlin, etc.

Zeitschrift für Pflanzenernährung,
und Bodenkunde. Weinheim

Zeitschrift für Pflanzenkrankheiten Pflanzenpathologie und
Pflanzenschutz. Stuttgart

Zeitschrift für Pflanzenphysiologie

Zeitschrift für Pflanzenzuchtung.
Berlin

Zeitschrift für Pilzkunde.
Heilbronn, Karlsruhe

Zeitschrift für Säugetierkunde
Berlin

Zeitschrift für Tierphysiologie,
Tierernährung und Futtermittel-
kunde. Hamburg

Zeitschrift für Tierpsychologie.
Berlin and Hamburg

Zeitschrift für Tierzüchtung und
Züchtungsbiologie. Berlin and
Hamburg

Zeitschrift für Tropenmedizin und
Parasitologie. Stuttgart

Zeitschrift für Urologie. Berlin
and Leipzig

Zeitschrift für Vererbungslehre.
Berlin, Heidelberg

Zeitschrift für vergleichende
Physiologie. Berlin

Zeitschrift für Versuchstierkunde.
Jena

Zeitschrift der Wiener entomologischen Gesellschaft. Wien

Zeitschrift für Wirtschaftsgeographie. Hagen, etc.

Zeitschrift für wissenschaftliche
Mikroskopie und für mikroskopische Technik. Leipzig, etc.

Zeitschrift für wissenschaftliche
Zoologie. Leipzig

Zeitschrift für Zellforschung und
mikroskopische anatomie. Berlin,
Wien

Zeitschrift für die Zucker-
industrie. Berlin

Zemědélská ekonomika. Československá akademie zemědélskych véd.
Praha

Zemědélská skola. Praha

Zemědélská technika. Sbornik
Ceskoslovenské akademie
zemědélské. Praha

Zemel'no-vodnye Resursy pustyn i
ikh Ispol'zovanie, Akademiya
nauk Turkmenskoi SSR. Ashkabad

Zement-Kalk-Gips. Wiesbaden

Zemizdat. Sofia

Zemledelie. Moskva

Zemljište i biljka. Beograd

Zentralblatt für allgemeine
Pathologie und pathologische
Anatomie. Jena

Zentralblatt für Bakteriologie,
Parasitenkunde, Infektions-
krankheiten und Hygiene. Jena,
Stuttgart

Zentralblatt für Chirurgie. Leipzig

Zentralblatt für Gynäkologie.
Leipzig

Zentralblatt für Veterinär-medizin.
Berlin, etc.

Zeszyty Badan Rejonow Uprzemys-
lawianych. Warszawa

Zeszyty Naukowe Politehniki
Slaskiej. Krakow

Zeszyty naukowe Szkoły głownej
gospodarstwa wiejskiego. Warszawa

Ser. Rolnictwo
Ser. Lesnictwo
Ser. Ogrodnictwo
Ser. Technologia

Zeszyty naukowe Towarzystwa
Naukowego Organizasji i
Kierownictiva. Warszawa

Zeszyty naukowe Uniwersytetu
łodzkiego. Łodz

Seria 2. Nauki matematyczno-
przyrodnicze

Zeszyty naukowe Uniwersytetu im.
Mikołaja Kopernika w Toruniu.
Toruń.

Zeszyty naukowe Wyższej szkoły
rolniczej. Weterynaria

Zeszyty naukowe Wyższej szkoły
rolniczej w Olsztynie. Warszawa

Zeszyty naukowe Wyższej szkoły
rolniczej w Krakowie. Krakow

Seria Rolnictwo

Zeszyty naukowe Wyższej szkoły
rolniczej w Szczecinie. Szczecin

Zeszyty naukowe Wyższej szkoły
rolniczej we Wrocławiu. Warszawa

Melioracja
Rolnictwo
Zootechnika

Zeszyty problemowe postępow nauk
rolniczych. Warszawa

Zhivotnovudni nauki. Bulgaria

Zhivotnovodstvo. USSR

Zhivotnovudstvo. Bulgaria

Zhurnal analiticheskoi khimii.
Moskva

Zhurnal fizicheskoi khimii.
Moskva

Zhurnal mikrobiologii. Kiev.

Zhurnal mikrobiologii, épidemio-
logii i immunobiologii. Moskva

Zhurnal nevropatologii i psik-
hiatrii imeni S.S. Korsakova.
Moskva

Zhurnal obshchei biologii.
Akademiya nauk SSSR. Moskva

Zhurnal Organicheskoi Khimii.
Moscow

Zhurnal prikladnoi khimii.
Moskva

Zhurnal Vsesoyuznogo Khimicheskogo
Obshchestva im D.I. Mendeleeva.
Moskva

Zhurnal Vysshei Nervnoi Deyatel'-
nosti imeni I.P. Pavlova. Moscow

Zierpflanzenbau.

Zivočišná výroba. Sbornik
Československé akademie
zemědélské. Praha

Zivotnovadstvo. Sofija

Zolfo in Agricultura. Palermo

Zoo. Société r. de zoologie
d'Anvers. Antwerp

Zooiatria. Revista de Medicina
Veterinaria y Produccion
Pecuaria. Chile

Zooleo. Société de botanique et de
zoologie congolaises.
Léopoldville

Zoologica. Scientific contributions of the New York
Zoological Society. New York

Zoologica. Original-Abhandlungen
aus dem Gesamtgebiete der
Zoologie. Stuttgart

Zoologica poloniae. Lwów, Wroclaw

Zoological Magazine. Tokyo

Zoological Record. London

   Section Insecta

Zoologicheskii sbornik. Akademiya nauk Armyanskoi SSR. Erevan

Zoologicheskii zhurnal. Moskva

Zoologiché listy. v Brné

Zoologische bijdragen. Rijks-museum van natuurlijke historie te Leiden. Leiden

Zoologische Garten. Leipzig

Zoologische Jahrbücher. Jena

   Allgemeine Zoologie und Physiologie der Tiere

   Anatomie und Ontogenie der Tiere

   Systematik, Ökologie und Geographie der Tiere

Zoölogische mededeelingen. Rijksmuseum van natuurlijke historie te Leiden. Leiden

Zoologische verhandelingen. Rijksmuseum van natuurlijke historie te Leiden. Leiden

Zoologischer Anzeiger. Leipzig

Zoologiska bidrag från Uppsala Uppsala och Stockholm

Zoology of Iceland. Copenhagen, Reykjavik

Zoology Publications from Victoria University College. Wellington, N.Z.

Zoonoses Research. New York

Zooprofilassi: rivisti mensile di scienza e tecnica veterinaria. Roma

Zootechnia. Societas internationalis veterinariorum zootechnicorum. Madrid

Zootecnia. São Paulo

Zootecnica e veterinaria: Fecondazione artificiale. Milano

Zuchthygiene. Berlin

Zuchtungskunde. Gottingen, etc.

Zucker. Hannover

Zuowu Xuebao

Zycie gospodarcze. Warszawa

Zycie Weterynaryjne. Poland